Local Jet Bundle Formulation
of
Bäcklund Transformations

MATHEMATICAL PHYSICS STUDIES

A SUPPLEMENTARY SERIES TO
LETTERS IN MATHEMATICAL PHYSICS

VOLUME 1

Local Jet Bundle Formulation of Bäcklund Transformations

With Applications to Non-Linear Evolution Equations

by

F. A. E. Pirani

D. C. Robinson

and

W. F. Shadwick

*Department of Mathematics, King's College,
University of London*

D. Reidel Publishing Company

Dordrecht : Holland / Boston : U.S.A.
London : England

Library of Congress Cataloging in Publication Data

Pirani, F. A. E. 1928–
    Local jet bundle formulation of Bäcklund transformations, with applications to non-linear
evolution equations.

    (Mathematical physics studies ; 1)
    Bibliography: p.
    Includes index.
    1. Differential equations, Partial.   2.   Bäcklund transformations.   3.   Jet bundles
(Mathematics)   I.   Robinson, David Clyde, 1942–   joint author.   II.   Shadwick, W. F.
1951–   joint author.   III.   Title.   IV.   Series.
QA374.P56   515′.353   79–19018
ISBN-13: 978-90-277-1036-9   e-ISBN-13: 978-94-009-9511-6
DOI: 10.1007/ 978-94-009-9511-6

Published by D. Reidel Publishing Company
P.O. Box 17, Dordrecht, Holland

Sold and distributed in the U.S.A., Canada, and Mexico
by D. Reidel Publishing Company, Inc.
Lincoln Building, 160 Old Derby Street, Hingham, Mass. 02043, U.S.A.

TABLE OF CONTENTS

ABSTRACT

Bäcklund transformations of systems of partial differential

equations are formulated in the language of jets. The aim of the

paper is not to present a complete and rigorous theory but to

establish a framework within which a variety of results concerning

non-linear evolution equations may be placed. The reader is not

assumed to have any previous familiarity with jets.

The central idea is the Bäcklund map, which is a smooth map

of jet bundles parametrized by the target manifold of its co-

domain. The original system of differential equations appears as

a system of integrability conditions for the Bäcklund map. The

map induces an horizontal distribution on its domain from the

natural contact structure of its codomain, which makes possible a

geometrical description in terms of a connection, called here the

Bäcklund connection; the system of integrability conditions

reappears as the vanishing of the curvature of this connection.

It is shown how $SL(2,\mathbb{R})$ and other groups which have been observed

to enter the theory of non-linear evolution equations may be

identified as structural groups of fibre bundles with Bäcklund

connections.

Prolongation of a Bäcklund map are defined; if the image of the prolonged integrability conditions under a prolonged Bäcklund map is a system of differential equations on its codomain then the correspondence between the original system and this image system is called a Bäcklund transformation.

A deformation of a Bäcklund map is defined and sufficient conditions are given for a deformation to leave the integrability conditions invariant. The problem of constructing Bäcklund maps is discussed.

The connection between this jet bundle formulation and related work, particularly that of Estabrook, Wahlquist and Hermann, is explained and various features of known examples are clarified.

Section 1

INTRODUCTION

The aim of this paper is to show that the theory of jet bundles
supplies the appropriate setting for the study of Bäcklund transformations. These transformations are used to solve certain
partial differential equations, particularly non-linear evolution
equations. Of course jets have been employed for some time in
the theory of partial differential equations, but so far little
use has been made of them in applications. In the meanwhile,
substantial progress has been made in the study of non-linear
evolution equations. This work has been encouraged by the dis-
covery of remarkable properties of some such equations, for
example the existence of soliton solutions and of infinite sequences of conservation laws. Among the techniques devised to
deal with these equations are the inverse scattering method and
the Bäcklund transformation. In our opinion the jet bundle
formulation offers a unifying geometrical framework for under-
standing the properties of non-linear evolution equations and
the techniques used to deal with them, although we do not consider
all of these properties and techniques here.

The relevance of the theory of jet bundles is that it

1

legitimates the practice of regarding the partial derivatives of
field variables as independent quantities. Since Bäcklund transformations require from the outset manipulation of these partial
derivatives, and repeated shifts of point of view about which
variables are dependent on which, this geometrical setting clari-
fies and simplifies the concepts involved, and offers the prospect
of bringing coherence to a variety of disparate results. Among
its specific advantages are

(1) that systems of partial differential equations with any
numbers of dependent and independent variables may be treated;

(2) that Bäcklund transformations of these systems admit
natural formulations;

(3) that it provides a suitable context for the discussion
of symmetries and conservation laws; and

(4) that it constitutes a basis for later global investigations.

The transformation devised by Bäcklund himself a century ago
[4,5] was used for the solution of a problem in Euclidean geometry
if, on a surface in Euclidean 3-space, the asymptotic directions
are taken as coordinate directions, then the surface metric may
be written in the form

$$ds^2 = dx^2 + 2 \cos z \, dx \, dy + dy^2 \quad , \tag{1.1}$$

where z is a function of the surface coordinates x and y. A
necessary and sufficient condition for the surface to be of

2

constant curvature -1 is that z satisfy the equation

$$\frac{\partial^2 z}{\partial x\, \partial y} = \sin z \ ,\qquad (1.2)$$

which is nowadays called the sine-Gordon equation. If z is a
known solution of this equation and if z' is another function of
x and y defined by the equations

$$\frac{\partial z'}{\partial x} = \frac{\partial z}{\partial x} + 2a\, \sin\tfrac{1}{2}(z' + z) \ ,$$

$$(1.3)$$

$$\frac{\partial z'}{\partial y} = -\frac{\partial z}{\partial y} + \frac{2}{a}\, \sin\tfrac{1}{2}(z' - z) \ ,$$

where a is a non-zero real parameter, then z' is also a solution
of the sine-Gordon equation. In fact, if the first of (1.3) is
differentiated with respect to y and the second with respect to
x, then z and its derivatives may be eliminated to yield the
sine-Gordon equation for z', or equally well, z' and its derivatives may be eliminated to yield the sine-Gordon equation for z.
Thus equations (1.3) may be used to construct new solutions of
the sine-Gordon equation from old. They have been known since
the work of Clairin [11] as a "Bäcklund transformation"; in this
paper, we distinguish between "Bäcklund maps", which are systems
of equations of which (1.3) is an example, and "Bäcklund transformations", which are correspondences between systems of partial
differential equations (in this case between an equation and
itself) defined by Bäcklund maps. More details concerning the
original geometrical problem and its solution will be found in a

book by Eisenhart [24].

Towards the end of the nineteenth century, transformations
of this kind, relating solutions of other partial differential
equations, were studied by Clairin [11] and by Goursat [33]. A
recent paper by Lamb [51] serves as an introduction to their
work. The common feature of these transformations is that a
certain function and its derivatives appear in first order partial
differential equations for another function, and that elimination
of one or the other yields a partial differential equation of
higher order which is either the equation of interest or a simpler
one thus related to it.

The recent period of application of these ideas goes back at
least to the work of Loewner [55], who employed them to solve
problems in continuum mechanics (see the survey by Rogers [64]).
Subsequently, the sine-Gordon equation has been employed to model
a wide variety of physical systems, and other equations have been
discovered to have similar properties and to admit similar transformations (see for example [9, 14, 16, 36, 72]). Outstanding
among these is the Korteweg-deVries (KdV) equation

$$\frac{\partial z}{\partial t} + 12z \frac{\partial z}{\partial x} + \frac{\partial^3 z}{\partial x^3} = 0 , \qquad (1.4)$$

which was originally derived before the turn of the century [49]
as an approximate description of small-amplitude water waves in
a narrow channel. It has been found to arise in a variety of
other physical situations [68] and to share with the sine-Gordon

4

equation the property of admitting multi-soliton solutions, which
break up asymptotically in time into solitary waves (for a
general review see the paper by Scott, Chu and McLaughlin [64]).
These exact solutions have now been obtained in a variety of
different ways [2, 28, 45, 71]. In particular, Wahlquist and
Estabrook [72] showed how to obtain them with the help of the
Bäcklund transformation

$$\frac{\partial y}{\partial x} = - 2z - y^2 + \lambda$$

$$(1.5)$$

$$\frac{\partial y}{\partial t} = 4(z + \lambda) (2z + y^2 - \lambda) + 2 \frac{\partial^2 z}{\partial x^2} - 4 \frac{\partial z}{\partial x} \cdot y$$

where λ is an arbitrary real parameter, which relates solutions
of the KdV equation to solutions of the equation

$$\frac{\partial y}{\partial t} + 6(\lambda - y^2) \frac{\partial y}{\partial x} + \frac{\partial^3 y}{\partial x^3} = 0 . \qquad (1.6)$$

Equations (1.5) may also be cast into a form suitable to the
application of the inverse scattering method, which was first
applied to the KdV equation by Gardner, Greene, Kruskal and
Miura [28, 29] and later developed in various ways by Lax [53],
Zakharov and Shabat [72], Ablowitz, Kaup, Newell and Segur [1,2],
and Calogero and Degasperis [7], among others.

This method converts the problem of solving equation (1.4)
into the problem of solving a linear integral equation. From
this point of view, the importance of the Bäcklund transformation
equations (1.5) is that they may be used to derive the equations

5

of the appropriate inverse scattering problem. The corresponding situation for the sine-Gordon equation is described below in example 5.1 (see also the paper by M. Crampin [15]).

To a great extent, the advantages of the jet bundle formulation listed above accrue also to the "prolongation" method of Wahlquist and Estabrook [72]. Their method is based on Cartan's theory of exterior differential systems [8]. It is remarkably fertile. When applied to a non-linear evolution equation it can yield the Bäcklund transformation, the linear equations of the associated inverse scattering problem, and an infinite-dimensional Lie algebra associated to the original equation. Our approach, which is stimulated by their work, is in some respects a generalization of theirs: we make use of the whole of the contact ideal on an appropriate jet bundle, while they employ only a sub-ideal. Their method requires fewer variables than ours, and is very efficient, while ours provides a more general framework within which the significance of specializations is more easily seen.

Hermann [43, 44], among others, has in effect pointed out that if the infinite-dimensional Lie algebra of Wahlquist and Estabrook is replaced by a finite-dimensional quotient, then their prolongation process may be interpreted as the construction of a connection, the curvature of which vanishes on solutions of the equation being prolonged. This connection appears very easily and fairly naturally in the jet bundle formulation.

To achieve our aim, we have included some didactic material

6

in this paper. Up to now the use of the theory of jet bundles in the study of partial differential equations has been pretty well confined to the mathematical, in contradistinction to the physical, literature. Therefore we have not assumed that the reader has any previous knowledge of jets, and in this respect we have attempted to write something reasonably self-contained. We have, on the other hand, assumed some familiarity with the elements of differential geometry, tensor calculus and exterior calculus. We have formulated many of our results, and some of our arguments, in coordinate-free terms, but have not attempted any global formulation. We remind the reader repeatedly that our investigations are all local, although we do not mention this on every possible occasion.

Numerous examples are treated in detail. In these examples we illustrate the general theory by reformulating or extending known results.

The notation and relevant facts of the theory of jet bundles are set out in section 2. Bäcklund maps and Bäcklund transformations are defined in sections 3 and 4. The relation to the theory of connections and their curvatures is described in section 5. Symmetry properties of Bäcklund maps are considered in section 6, and the determination of Bäcklund maps for a given system of partial differential equations in section 7, where our method is compared with that of Wahlquist and Estabrook. The latter section leaves open a number of important questions concerning the existence of Bäcklund maps.

We conclude this introduction with a more detailed outline of the main mathematical ideas of the paper. The reader unfamiliar with jets will not lose continuity if he now skips to the last paragraph of this section where we explain the notation to be used in the following sections.

In section 2 we introduce contact structures and give a definition of systems of partial differential equations sufficiently general for our needs: a system of partial differential equations of order k means a submanifold of a k-jet bundle $J^k(M,N)$ whose source M and target N are the manifolds of independent and dependent variables respectively. It is assumed that this submanifold is the zero set Z of a finitely-generated ideal of functions on $J^k(M,N)$.

We next introduce the central ideas of the paper [63], treating second order systems of partial differential equations in section 3, and systems of arbitrary order in section 4. In the general case, for example, we construct a map

$$\psi : J^h(M,N_1) \times N_2 \to J^1(M,N_2) , \qquad (1.7)$$

where M is the manifold of independent variables and N_1 and N_2 are the manifolds of old and new dependent variables respectively (in equations (1.3), which comprise an instance of (1.7), x and y correspond to coordinates on M, z and z′ to coordinates on N_1 and N_2 respectively). If the integrability conditions for (1.7) comprise a system $\tilde{Z}$ of (parametrized) differential equations on $J^{h+1}(M,N_1) \times N_2$ then we call ψ a Bäcklund map. If, as is generally the case in applications, there is a system of differential equations Z on $J^{h+1}(M,N_1)$ such that $Z \times N_2 \subset \tilde{Z}$, then we call ψ an ordinary Bäcklund map for Z. We define the prolongations ψ^s and Z^s of ψ and Z, and if there is a least integer s such that the image of $\psi^s \big|_{Z^s}$ is a system Z' of differential equations on $J^{s+1}(M,N_2)$ then we call the correspondence between Z and Z' the Bäcklund transformation determined by ψ. If a system Z of partial differential equations is given on $J^{h+1}(M,N_1)$, then we refer to the determination of manifolds N_2 and maps (1.7) which are ordinary Bäcklund maps for Z as the Bäcklund problem for Z.

In Section 5 we show that a Bäcklund map ψ naturally determines a connection $H_{f,\psi}$ on $J^0(M,N_1) \to M$, associated to any function $f : M \to N_1$, and show that the curvature of this connection vanishes if and only if f satisfies the integrability conditions for ψ. We remark that ψ also naturally determines a connection H_ψ on $J(M,N_1) \times N_2 \to M$, where $J(M,N_1)$ denotes the jet bundle of infinite order. In this case the curvature vanishes on the integrability conditions for ψ themselves. If there is a Lie group G such that $H_{f,\psi}$ is a G-connection – a circumstance which is independent of the choice of f – then there is a solution of the Bäcklund problem determining a linear G-connection, and in the application to certain non-linear evolution equations, the associated linear scattering equations are the equations of

parallel transport determined by this linear connection. The
soliton connection previously introduced [18] is the connection
on a principal bundle associated to the bundle with bundle space
$J^O(M,N_1)$.

In section 6 we investigate extended point transformations
of jet bundles compatible with a given Bäcklund map. We generalize the well-known result that the Bäcklund map for the sine-
Gordon equation is composed of (Lie) transformations of the
independent variables and a parameter-free (Bianchi) map [54],
and observe that the eigenvalue in the linear scattering equations
associated with certain non-linear evolution equations may be
identified with the parameter in a 1-parameter group of extended
point transformations.

In section 7, besides discussing the solution of Bäcklund
problems, and comparing our method with that of Wahlquist and
Estabrook [25, 72], which we generalise to deal with an arbitrary
number of independent variables, we also consider their procedure
for relating two different solutions of a given Bäcklund problem,
and identify the Lie algebra obtained by them in prolonging the
Korteweg-deVries equation.

We conclude this introduction with a summary of our notation
and conventions. All objects and maps are assumed to be C^∞; in
the applications they are generally real-analytic. If $f : M \rightarrow N$
is a map, then the domain of f is an open set in M, not necessarily
the whole of M. The range and summation conventions employed are

stated explicitly when first they are assumed. Where a coordinate appears as an argument of the coordinate presentation of a map, the index is understood to range but not sum. If Q is any set, then $\Delta(Q)$ denotes the diagonal map $Q \to Q \times Q$ by $q \mapsto (q,q)$, for every $q \in Q$. If ϕ is a map of manifolds, then ϕ^* is the induced map of forms and functions. If Ω is any collection of exterior forms then $\phi^*\Omega$ means $\{\phi^* \omega | \omega \in \Omega\}$, $d\Omega$ means $\{d\omega | \omega \in \Omega\}$, $I(\Omega)$ means the ideal $\{\Sigma \eta \wedge \omega | \omega \in \Omega, \eta \text{ any forms}\}$ generated by Ω, where $\wedge$ denotes the exterior product, and $X \lrcorner \Omega$ means $\{X \lrcorner \omega | \omega \in \Omega\}$ where X is any vector field and $\lrcorner$ denotes the interior product of a vector field and a form (see for example [39]). If X is a vector field, the Lie derivative along X is denoted $\pounds_X$. Projection of a Cartesian product on the ith factor is denoted pr_i. The bracket of vector fields is denoted $[\ ,\ ]$. The end of an example is denoted $\square$.

Acknowledgements: We have benefited greatly from conversations with J. Corones, M. Crampin, P.F.J. Dhooghe, R. Dodd, B. Fox, R. Hermann, L.H. Hodgkin, H. Levy, P.J. McCarthy, H.C. Morris, A. Solomonides, H. Wahlquist and T.J. Willmore. This work was partly supported by a grant from the Norman Foundation. The third author (W.F.S.) is supported by a National Research Council of Canada Scholarship. Mrs. J.L. Bunn patiently typed and retyped the manuscript.

Section 2

JET BUNDLES

In this section we summarize the notation and relevant facts of
the theory of jet bundles. Our summary is not intended to be
complete, but only to explain those parts of the theory which we
need later. As we explained in section 1, this theory is useful
for the discussion of Bäcklund transformations because it provides a rigorous basis for the manipulation of partial derivatives
as if they were independent variables. The formulation of the
theory is due to C. Ehresmann [23]. There are brief mathematical
introductions in the book by Golubitsky and Guillemin [35] and
in the memoir by Guillemin and Sternberg [35]. More leisurely
presentations may be found in several books by R. Hermann [40,
41, 42]. Jet bundles have often been used before in the study of
partial differential equations, for example by H.H. Johnson [48],
H. Goldschmidt [31, 32], D. Krupka and A. Trautman [50, 51, 69].
The applications we have in mind are rather different from
theirs, but we should mention that the idea of a contact module,
introduced in this section, is taken from Johnson's paper, and
the definition of a prolongation imitated from Goldschmidt's.

Let M and N be C^∞ manifolds, and let $C^\infty(M,N)$ denote the

12

collection of C^∞ maps f: U → N, where U is an open set in M

(recall that our considerations are all local). In the

applications,

M is a space of independent variables, and

N is a space of dependent variables.

Two maps f, g $\in$ C^∞(M,N) are said to agree to order k at

x $\in$ M if there are coordinate charts around x $\in$ M and around

f(x) = g(x) $\in$ N in which they have the same Taylor expansion up

to and including order k. It is easy to convince oneself that if

f and g agree to order k for one choice of coordinate charts then

they agree to order k for any other choice, so that agreement to

order k is coordinate-independent. It is also an equivalence

relation: the equivalence class of maps which agree with f to

order k at x is called the k-jet of f at x and denoted $j^k_x f$.

If x^a are local coordinates around x $\in$ M and z^μ are local

coordinates around f(x) $\in$ N, then $j^k_x f$ is determined by the

numbers

$$x^a, \quad z^\mu = f^\mu(x), \quad z^\mu_a = \partial_a f^\mu(x),$$

$$z^\mu_{ab} = \partial_{ab} f^\mu(x), \quad \ldots, \quad z^\mu_{a_1 \ldots a_k} = \partial_{a_1 \ldots a_k} f^\mu(x),$$

where $f^\mu(x)$ is the coordinate presentation of f, and (here and

throughout) $\partial_a, \ldots, \partial_{a_1 \ldots a_k}$ denote partial derivatives:

$$\partial_a f^\mu := \frac{\partial f^\mu}{\partial x^a}, \quad \partial_{a_1 \ldots a_k} f^\mu := \frac{\partial^k f^\mu}{\partial x^{a_1} \ldots \partial x^{a_k}} \ .$$

Here and below, lower case Latin indices a, b, a_1, a_2 ... range

and sum over 1, ..., dim M and Greek indices μ, ν, ... range and
sum over 1, ..., dim N. However, the lower case Latin indices
h, k, ℓ, m, q and s denote non-negative integers, and are not
subject to the range or summation conventions.

Conversely, any collection of numbers

$$x^a, \; z^\mu, \; z^\mu_{\;a}, \; z^\mu_{\;ab}, \; \ldots, \; z^\mu_{\;a_1\ldots a_k} \;\; ,$$

with x^a and z^μ in the coordinate ranges of the corresponding
charts, and $z^\mu_{\;ab}, \; \ldots, \; z^\mu_{\;a_1\ldots a_k}$ symmetric in their lower indices,
but all the numbers otherwise arbitrary, determine a unique
equivalence class. Putting this in a more formal setting, one
obtains the definition of the k-jet bundle.

The k-jet bundle of M and N, denoted $J^k(M,N)$, is the set of
all k-jets $j^k_x f$, with k fixed, $x \in M$, and $f \in C^\infty(M,N)$. It may
be shown to have a natural differentiable structure ([3] section
15). The map

$$\alpha : J^k(M,N) \to M$$

by $\quad j^k_x f \mapsto x$

is called the source map, and x is called the source of $j^k_x f$.
The map

$$\beta : J^k(M,N) \to N$$

by $\quad j^k_x f \mapsto f(x)$

is called the target map, and $f(x)$ is called the target of $j_x^k f$.
A point $\xi \in J^k(M,N)$ is an equivalence class of maps with

the same source,

the same target, and

the same derivatives up to the kth in every coordinate

presentation.

Now let x^a be local coordinates around $\alpha(\xi) \in M$ and let z^μ

be local coordinates around $\beta(\xi) \in N$. Let z^μ_a, z^μ_{ab}, ...,

$z^\mu_{a_1 \ldots a_k}$ denote the derivatives, up to the kth, of the co-

ordinate presentation of any map in the equivalence class ξ. By

a common abuse, write x^a and z^μ for the induced functions $\alpha^* x^a$

and $\beta^* z^\mu$. Then x^a, z^μ, z^μ_a, z^μ_{ab}, ..., $z^\mu_{a_1 \ldots a_k}$ may be chosen

as local coordinates around ξ. They will be called the standard

coordinates on $J^k(M,N)$, associated to the coordinates x^a on M and

z^μ on N. With this convention, the standard coordinates on

$J^\ell(M,N)$, for any $\ell \neq k$, are x^a, z^μ, z^μ_a, z^μ_{ab}, ... $z^\mu_{a_1 \ldots a_\ell}$,

but this ambiguity causes no difficulties.

If $k > \ell$, then ignoring all derivatives above the ℓth

yields the natural projection of the k-jet bundle on the ℓ-jet

bundle:

$$\pi^k_\ell : J^k(M,N) \to J^\ell(M,N)$$

by $\quad j_x^k f \to j_x^\ell f$.

In standard coordinates,

$$\pi^k_\ell(x^a, z^\mu, z^\mu_a, z^\mu_{ab}, \ldots, z^\mu_{a_1 \ldots a_\ell}, z^\mu_{a_1 \ldots a_{\ell+1}}, \ldots, z^\mu_{a_1 \ldots a_k})$$

$$= (x^a, z^\mu, z^\mu_a, z^\mu_{ab}, \ldots z^\mu_{a_1 \ldots a_\ell}).$$

In particular, $J^O(M,N)$ may be identified with $M \times N$, and

$$\pi^k_O = (\alpha \times \beta) \circ \Delta(J^k(M,N)).$$

Moreover, π^ℓ_ℓ is understood to mean the identity map of $J^\ell(M,N)$.

An equivalent definition of the k-jet bundle may be formulated, without reference to coordinates, in terms of tangent maps, as is evident from the following proposition [3] : Let $T\phi$ denote the tangent map to any map of manifolds $\phi : M \to N$. Let $T^1\phi := T\phi$ and $T^k\phi = T(T^{k-1}\phi)$. Then f, g ϵ $C^\infty(M,N)$ agree to order k at x ϵ M if and only if $T^k f\big|_x = T^k g\big|_x$.

Any given map determines a k-jet at each point of its domain. Thus if f ϵ $C^\infty(M,N)$, the k-jet extension of f is

$$j^k f : U \to j^k(M,N)$$

by $\quad x \mapsto j^k_x f$,

where U is the domain of f.

The k-jet extension of f is a cross-section of the source map α:

$$\alpha \circ j^k f = \mathrm{id}_U .$$

In the case k = 0, $j^O f$ is just the graph of f.

A map from a jet bundle to another manifold induces maps of

higher jet bundles, called prolongations, which in coordinates amount merely to the taking of total derivatives. The construction is as follows: let M, N and P be manifolds and let

$$\phi : J^h(M,N) \to P$$

be a smooth map. The sth prolongation of ϕ is the unique map

$$p^s\phi : J^{h+s}(M,N) \to J^s(M,P)$$

with the property that for every $f \in C^\infty(M,N)$, the diagram

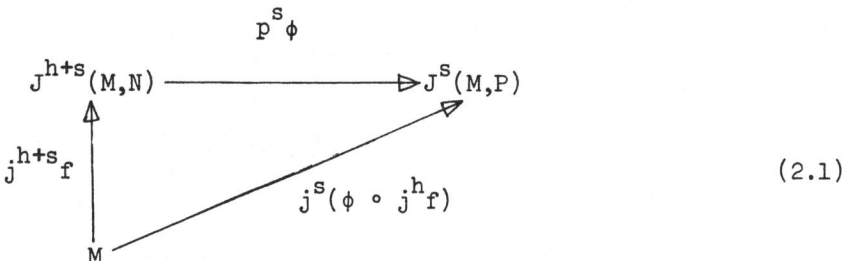

(2.1)

commutes. If x^a, z^μ and y^A are local coordinates on M, N and P respectively, with the above conventions on a and μ, and A ranging over 1, ..., dim P, and if in standard coordinates ϕ has the presentation

$$y^A = \phi^A(x^a, z^\mu, z^\mu_a, \ldots, z^\mu_{a_1 \ldots a_h})$$

and f has the presentation

$$z^\mu = f^\mu(x^a),$$

then $\phi \circ j^h f$ has the presentation

17

$$y^A = \phi^A(x^a, f^\mu(x), \partial_a f^\mu(x), \ldots, \partial_{a_1 \ldots a_h} f^\mu(x)).$$

Now let $D^{(k)}_b$ denote the total derivative:

$$D^{(k)}_b := \frac{\partial}{\partial x^b} + z^\mu_{\ b} \frac{\partial}{\partial z^\mu} + \ldots + z^\mu_{\ bb_1 \ldots b_{k-1}} \frac{\partial}{\partial z^\mu_{\ b_1 \ldots b_{k-1}}}. \qquad (2.2)$$

It is a coordinate-dependent collection of (dim M) vector fields, defined on $J^\ell(M,N)$ for $\ell \geq k$. Its important property is that for any function ϕ of the indicated arguments (and any $k > h$),

$$\frac{\partial}{\partial x^b} [\phi(x^a, f^\mu(x), \partial_a f^\mu(x), \ldots, \partial_{a_1 \ldots a_h} f^\mu(x))]$$

$$= [D^{(k)}_b \pi^{h+1*}_h \phi] (x^a, f^\mu(x), \partial_a f^\mu(x), \ldots, \partial_{a_1 \ldots a_{h+1}} f^\mu(x)),$$

which is to say that the same result is achieved either by first evaluating ϕ on f and its derivatives, and then taking the partial derivative, or by first taking the total derivative of ϕ and then evaluating on f and its derivatives. Some other properties are enumerated below, starting with equation (2.9).

For clarity we introduce two abbreviations:

(1) If $D^{(k)}_b$ acts on a function, it is assumed that the function, if not defined on $J^k(M,N)$, has been lifted to $J^k(M,N)$ with the natural projection. Thus

$$D^{(k)}_b \pi^{k*}_\ell \text{ will be abbreviated } D^{(k)}_b.$$

(2) Repeated total differentiation will be denoted $D^{(k)}_{b_1 \ldots b_s}$, namely

$D^{(k)}{}_{b_1} \; D^{(k)}{}_{b_2} \; \cdots \; D^{(k)}{}_{b_s}$ will be abbreviated $D^{(k)}{}_{b_1 \ldots b_s}$.

With this notation, the coordinate presentation of $j^s(\phi \circ j^h f)$ is

$$y^A = \phi^A(x^a, \; f^\mu(x), \; \partial_a f^\mu(x), \ldots, \; \partial_{a_1 \ldots a_h} f^\mu(x))$$

$$y^A{}_b = [D^{(h+s)}{}_b \phi^A](x^a, \; f^\mu(x), \; \partial_a f^\mu(x), \; \ldots, \; \partial_{a_1 \ldots a_{h+1}} f^\mu(x))$$

$$\vdots \tag{2.3}$$

$$y^A{}_{b_1 \ldots b_s} = [D^{(h+s)}{}_{b_1 \ldots b_s} \phi^A](x^a, \; f^\mu(x), \; \ldots, \; \partial_{a_1 \ldots a_{h+s}} f^\mu(x)).$$

Since (2.3) must hold for every f, $p^s \phi$ must have the presentation

$$x^a = x^a$$

$$y^A = \phi^A(x^a, \; z^\mu, \; z^\mu{}_a, \; \ldots, \; z^\mu{}_{a_1 \ldots a_h})$$

$$y^A{}_b = [D^{(h+s)}{}_b \; \phi^A] \; (x^a, \; z^\mu, \; z^\mu{}_a, \; \ldots, \; z^\mu{}_{a_1 \ldots a_{h+1}})$$

$$\vdots \tag{2.4}$$

$$y^A{}_{b_1 \ldots b_s} = [D^{(h+s)}{}_{b_1 \ldots b_s} \; \phi^A] \; (x^a, \; z^\mu, \; z^\mu{}_a, \ldots, \; z^\mu{}_{a_1 \ldots a_{h+s}}).$$

Thus the map $p^s \phi$ comprises ϕ itself and its first s total derivatives with respect to the coordinates on M.

The question, when is a map a k-jet, which arises frequently, may be answered neatly with the help of contact modules, which are naturally-defined structures on jet bundles. We introduce them with the help of a simple classical example.

19

<u>Example 2.1:</u> <u>Contact structure on a 1-jet bundle.</u> Let $M = \mathbb{R}^2$,

$N = \mathbb{R}^1$, and consider the 1-jet bundle $J^1(M,N)$. Let x^1 and x^2 be

coordinates on M and z^1 the coordinate on N. Then standard co-

ordinates on $J^1(M,N)$ are x^1, x^2, z^1, $z^1_{\ 1}$ and $z^1_{\ 2}$. In the class-

ical literature of partial differential equations these coordinates

are denoted x, y, z, p and q respectively. Now let $f \in C^\infty(M,N)$.

The 1-jet extension of f is

$$j^1f : M \to J^1(M,N)$$

with coordinate presentation

$$(x^1,x^2) \mapsto (x^1,x^2,z^1 = f(x^1,x^2), \ z^1_{\ 1} = \partial_1 f(x^1,x^2), \ z^1_{\ 2} = \partial_2 f(x^1,x^2)).$$

Now consider the 1-form θ^1 which in the classical notation is

$$\theta^1 = dz - p \ dx - q \ dy \qquad\qquad\qquad (2.5)$$

and in the standard coordinates defined above is

$$\theta^1 = dz^1 - z^1_{\ a} \ dx^a.$$

The map j^1f pulls forms back from $J^1(M,N)$ to M. In particular

$$(j^1f)^*\theta^1 = df(x^1,x^2) - \frac{\partial f}{\partial x^1}(x^1,x^2) \ dx^1 - \frac{\partial f}{\partial x^2}(x^1,x^2) \ dx^2 = 0$$

for any f. This says simply that the independent variables $z^1_{\ 1}$

and $z^1_{\ 2}$ on $J^1(M,N)$ become partial derivatives of z^1 when they are

evaluated on any 1-jet extension. It may be shown, conversely,

that if $F : M \to J^1(M,N)$ is a cross-section of the source map α

and if $F^*\theta^1 = 0$ then F is locally the 1-jet extension of a map

from M to N ([48] p.313).   □

The idea expressed in this example is easily generalized to

an arbitrary jet bundle. A contact form θ on $J^k(M,N)$ is a

1-form with the property

$$(j^k f)^* \theta = 0 \tag{2.6}$$

for every $f \in C^\infty(M,N)$. If θ_1 and θ_2 are contact forms, so is

$u_1\theta_1 + u_2\theta_2$, where u_1 and u_2 are arbitrary functions on $J^k(M,N)$;

thus contact forms comprise a module ([62] p.1) over

$C^\infty(J^k(M,N),\ \mathbb{R})$. This module will be denoted $\Omega^k(M,N)$, or, if no

ambiguity arises, Ω^k and called the contact module of $J^k(M,N)$

[48]. If $k > \ell$, then $\pi_\ell^{k*}\ \Omega^\ell \subset \Omega^k$, so that Ω^ℓ may be considered

to be a submodule of Ω^k.

Each contact module is finitely generated; in fact Ω^k has

a basis comprising the forms given in standard coordinates by

$$\theta^\mu = dz^\mu - z^\mu_{\ c}\ dx^c,$$

$$\theta^\mu_{\ b} = dz^\mu_{\ b} - z^\mu_{\ bc}\ dx^c,$$

$$\vdots \tag{2.7}$$

$$\theta^\mu_{\ b_1 \ldots b_{k-1}} = dz^\mu_{\ b_1 \ldots b_{k-1}} - z^\mu_{\ b_1 \ldots b_{k-1} c}\ dx^c.$$

This basis will be called the standard basis for Ω^k. It is

easy to see that if $f \in C^\infty(M,N)$ then $(j^k f)^*$ annihilates all these

21

forms precisely because it evaluates partial derivatives, and the assertion that (2.7) is a basis for the contact module says that there are no others linearly independent of these (for a proof see [62]).

In many applications it is necessary to know whether an ideal of differential forms is closed under exterior differentiation. An ideal $I(\Omega)$ is called closed, or a differential ideal, if $I(d\Omega) \subset I(\Omega)$.

In the present case, observe that

$$d\theta^\mu = dx^b \wedge \theta^\mu{}_b ,$$

$$d\theta^\mu{}_{b_1} = dx^b \wedge \theta^\mu{}_{bb_1} ,$$

$$\vdots \tag{2.8}$$

$$d\theta^\mu{}_{b_1 \cdots b_{k-2}} = dx^b \wedge \theta^\mu{}_{bb_1 \cdots b_{k-2}}$$

but that $d\theta^\mu{}_{bb_1 \cdots b_{k-2}}$ is not a linear combination of the forms in Ω^k. Thus $I(\Omega^k)$ is not closed. This is one of several related inconveniences which result from working on $J^k(M,N)$ with finite k. These inconveniences are formal manifestations of the fact that differentiation of a kth derivative yields a $(k + 1)^{st}$ derivative. They may be avoided by working on the jet bundle of infinite order $J(M,N)$ to be defined in section 5. For most of the applications considered here it seems enough to confine oneself to jet bundles of finite order.

22

For later use we record some easily verified properties of total derivative operators and contact ideals:

1. If u is any function on $J^{k-1}(M,N)$, then

$$d(\pi_{k-1}^{k*}u) = D_a^{(k)}u \ dx^a \quad \text{mod } \Omega^k , \qquad (2.9)$$

where "mod Ω^k" means that forms lying in Ω^k are discarded. This equation may be used as a definition of $D_a^{(k)}$, the local charts being given.

2. $$[D_a^{(k)} , D_b^{(k)}] = 0 \qquad (2.10)$$

where [,] denotes the bracket of vector fields.

3. If X is a vector (field) of the form

$$X = X^a \ D_a^{(k)} ,$$

then $X \lrcorner \Omega^k = 0$, $\qquad (2.11)$

but this is not enough to characterise $D_a^{(k)}$, because in addition

$$\frac{\partial}{\partial z^\mu_{b_1 \ldots b_k}} \lrcorner \Omega^k = 0 .$$

4. $$D_a^{(k)} \lrcorner d\theta^\mu = \theta^\mu_a,$$

$$D_a^{(k)} \lrcorner d\theta^\mu_{b_1} = \theta^\mu_{ab_1} , \qquad (2.12)$$

$$\vdots$$

$$D_a^{(k)} \lrcorner d\theta^\mu_{b_1 \ldots b_{k-2}} = \theta^\mu_{ab_1 \ldots b_{k-2}} .$$

In the language of jet bundles, a differential equation defines a subset of a jet bundle. For our purposes, which are not concerned with singularities of equations or of their solutions, it is convenient to define a system of partial differential equations of order k to be a submanifold Z of $J^k(M,N)$ which is the zero set of a finitely generated ideal Σ_Z of functions on $J^k(M,N)$. If F_1, F_2, ..., F_q is a set of generators, then the system of equations may be written in terms of the coordinate presentations of these functions - say, in standard coordinates

$$F_1(x^a, z^\mu, z^\mu{}_b, \ldots, z^\mu{}_{b_1 \ldots b_k}) = 0,$$

$$\vdots \qquad\qquad\qquad\qquad\qquad (2.13)$$

$$F_q(x^a, z^\mu, z^\mu{}_b, \ldots, z^\mu{}_{b_1 \ldots b_k}) = 0.$$

Thus the generators define a map

$$F : J^k(M,N) \to \mathbb{R}^q$$

and a solution of the system is a map $f \in C^\infty(M,N)$ such that

$$F \circ j^k f = 0 .$$

In the language ordinarily used in applications, one would refer to equations (2.13) themselves, rather than to the corresponding zero set, as "the system of differential equations". We shall use the phrase in this sense as well, making it clear in context which sense is meant.

24

Taking total derivatives of the generating functions $F_1 \ldots F_q$, one obtains equations of higher order which are consequences of the given ones. In the language of prolongations introduced above, the sth prolongation is the zero set Z^s of

$$p^s \ F : \ J^{k+s}(M,N) \rightarrow J^s(M,\mathbb{R}^q)$$

generated by all the functions with presentation F_α, $D^{(k+s)}_{b_1} F_\alpha$, $\ldots$, $D^{(k+s)}_{b_1 \ldots b_s} F_\alpha$ (where F_α is written, abusively, for $\pi^{k+s*}_k F_\alpha$).

We conclude this section with some examples which will also be useful later on:

Example 2.2: The sine-Gordon equation. As an example of the application of this formalism, consider the sine-Gordon equation (1.2), repeated here for convenience:

$$\frac{\partial^2 z}{\partial x \, \partial y} = \sin z \ . \tag{2.14}$$

In the notation of this section, since there are two independent variables and one dependent variable, let dim M = 2, and dim N = 1, writing x^1 and x^2 for x and y. To follow the notation systematically, one should also write z^1 for z, but from now on, to reduce the cloud of indices, we shall omit the index when 1 is the only value which it can take. Therefore, here and below, we write z_a, z_{ab} and so on instead of z^1_a, z^1_{ab} , $\ldots$. Since the equation is of second order it defines a subset of $J^2(M,N)$ which is the zero set Z of the single function

$$F = z_{12} - \sin z .\qquad\qquad (2.15)$$

The first prolongation of Z, on $J^3(M,N)$, is generated by

$$z_{12} - \sin z, \quad z_{112} - z_1 \cos z, \quad z_{122} - z_2 \cos z.$$

The contact module Ω^2 is generated by $\theta = dz - z_1 dx^1 - z_2 dx^2$ and by $\theta_1 = dz_1 - z_{11} dx^1 - z_{12} dx^2$ and $\theta_2 = dz_2 - z_{21} dx^1 - z_{22} dx^2$. In classical notation the last two would be written

$$\theta_1 = dp - r\ dx - s\ dy \text{ and } \theta_2 = dq - s\ dx - t\ dy.$$

The function

$$f : M \to N$$

by

$$(x^1, x^2) \mapsto z = f(x^1, x^2)$$

is a solution of the sine-Gordon equation if

$$F \circ j^2 f = 0$$

which (compare example 2.1) means that

$$\frac{\partial^2 f}{\partial x^1\ \partial x^2} (x^1, x^2) - \sin(f(x^1, x^2)) = 0 ,$$

which is of course just (2.14) all over again. $\qquad\qquad \square$

Example 2.3.   The Korteweg-deVries equation. This is equation (1.4), in classical notation

$$\frac{\partial z}{\partial y} + 12 z \frac{\partial z}{\partial x} + \frac{\partial^3 z}{\partial x^3} = 0 .\qquad\qquad (2.16)$$

In the notation of this section, again with dim $M = 2$, dim $N = 1$,

26

the equation is the zero set of the single function

$$F = z_2 + 12zz_1 + z_{111} \tag{2.17}$$

on $J^3(M,N)$, and the contact ideal is generated by

$$\theta = dz - z_a \, dx^a. \qquad \theta_a = dz_a - z_{ab} \, dx^b, \quad \text{and}$$

$$\theta_{ab} = dz_{ab} - z_{abc} \, dx^c. \qquad\qquad \square$$

Example 2.4. Contact transformations. In the classical theory, a contact transformation is a change of variables preserving $dz - p \, dx - q \, dy$ up to a factor; here it may be described as a (local) diffeomorphism ϕ of $J^1(M,N)$ satisfying

$$\phi^* \, \Omega^1 = \Omega^1 \quad . \tag{2.18}$$

It follows from the remark at the end of Example 2.1 that if Z is a submanifold corresponding to a system of differential equations and if $\phi(Z) \subset Z$ then $\phi \circ j^1 f$ is the 1-jet extension of a solution of Z whenever f is a solution of Z.

In the classical situation, where dim M = 2, dim N = 1, as in the preceding examples, there is a canonical construction of a vector field from a 1-form on $J^1(M,N)$, which may be extended to the case where dim M is arbitrary, dim N = 1. Given the contact form

$$\theta = dz - z_b \, dx^b$$

and any 1-form ω, define the vector field X on $J^1(M,N)$ by

$$X \lrcorner \theta = 0 \; ,$$

and $X \lrcorner d\theta - \omega \in \Omega^1$

or equivalently $(X \lrcorner d\theta - \omega) \wedge \theta = 0.$

In particular if $\omega = df$ write X_f for X. Then functions on $J^1(M,N)$ may be given a Lie algebra structure by defining

$$[f,g] = X_f g$$

for f, $g \in C^\infty(J^1(M,N), \mathbb{R})$. It is not difficult to show that if ϕ is a contact transformation and if

$$\phi^* \theta = \sigma \cdot \theta$$

where σ is a function on $J^1(M,N)$,

then $[\phi^* f, \phi^* g] = \sigma \cdot \phi^* [f,g]$,

so that ϕ is an automorphism of the Lie algebra structure.

An obvious generalization of contact transformations would be to (local) diffeomorphisms of $J^k(M,N)$ which preserve Ω^k. However, for finite k, any diffeomorphism which preserves Ω^k preserves Ω^1, and is actually (the prolongation of) a contact transformation [30,46]. Moreover, if dim N > 1, such a diffeomorphism must be the prolongation of a transformation of $J^0(M,N) \simeq M \times N$ [46].

This limitation to contact transformations may be escaped by considering the jet bundle of infinite order, to which the preceding argument does not apply (compare section 5). On the other hand one may take the view that a Bäcklund map is a kind of generalization of a contact transformation, parametrized with the new independent variables, which also escapes this limitation (compare [46] and section 6 below).

Note

1. It is assumed throughout that a differential equation and its prolongations are such that the maps $\alpha|_Z$ and $\alpha|_Z$s are surjective submersions.

Section 3

BÄCKLUND MAPS : SIMPLEST CASE

A Bäcklund map is a transformation of the dependent variables in
a system of differential equations, whereby the first derivatives
of the new variables are given in terms of the new variables
themselves as well as of the old variables and their derivatives.
In this section, only the first derivatives of the old variables
enter in. The general case, where higher derivatives are
admitted, is discussed in the next section.

In the literature, Bäcklund maps are usually called "Bäcklund
transformations". However "Bäcklund transformation" is also used
to mean the correspondence between systems of differential
equations induced by a Bäcklund map. Here we shall reserve
"Bäcklund transformation" to mean this correspondence only.

In this paper, the independent variables are left unaltered,
although in some applications one might wish to drop this restriction. The presentation which follows generalizes much of
the summary by Forsyth [26] of the version published earlier by
Clairin [11]. As Clairin remarks, reading the original papers
by Bäcklund is "malheureusement un peu difficile".

Let M, N_1 and N_2 be C^∞ manifolds, and let

30

$$\psi : J^1(M, N_1) \times N_2 \rightarrow J^1(M, N_2) \tag{3.1}$$

be a C^∞ map. In the applications,

M is the space of the independent variables,

N_1 is the space of the old dependent variables, and

N_2 is the space of the new dependent variables.

In this section we shall put conditions on ψ which are appropriate for it to be called a "Bäcklund map", and explain their significance. In the classical examples, and some others, N_1 and N_2 are diffeomorphic, which for our local considerations means merely that the number of new variables is the same as the number of old ones, but in general this should not be supposed to be the case.

The first condition on ψ is that the new dependent variables – the coordinates on N_2 – should be unaltered by the map. This will be the case if ψ acts trivially on N_2, that is, if

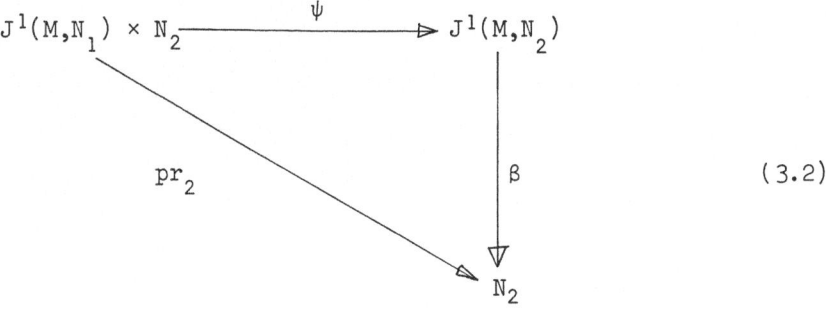

$$\tag{3.2}$$

commutes. Here, as in section 2, β denotes the target map.

As stated above, in this paper it is supposed that the independent variables – the coordinates on M – are also left unaltered by the map ψ, which will be the case if ψ acts trivially on M, that is, if

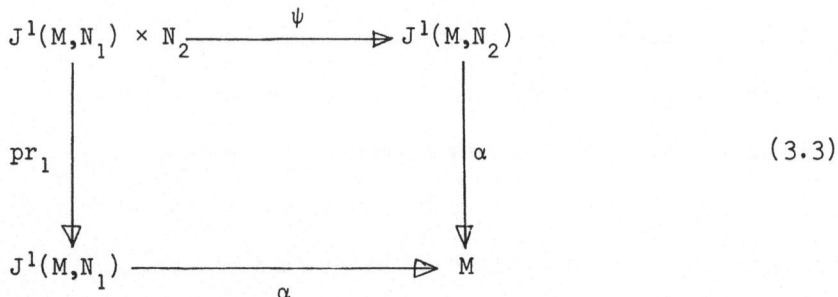

$$(3.3)$$

commutes. Here pr_1 means projection on the first factor, and, as in section 2, α denotes the source map.

These two assumptions about ψ leave only the first derivatives of the new dependent variables still to be determined. This becomes apparent when one introduces local coordinates:
On the domain of ψ, choose local coordinates

x^a on M, $a,b,c, \ldots = 1, \ldots,$ dim M

z^μ on N_1, $\mu,\rho, \ldots = 1, \ldots,$ dim N_1

y^A on N_2, $A,B,C, \ldots = 1, \ldots,$ dim N_2

and standard associated coordinates z^μ_b on $J^1(M,N_1)$ (with range and summation conventions for each alphabet).
On the codomain of ψ, choose local coordinates

x'^a on M,

y'^A on N_2,

and standard associated coordinates y'^A_b on $J^1(M,N_2)$.
Then the commuting of (3.2) may be expressed

$$y'^A = y^A .$$

$$(3.4)$$

This will always be assumed to hold, and if no ambiguity arises y^A will be written, abusively, in place of y'^A.

The commuting of (3.3) may be expressed

$$x'^a = x^a.$$ (3.5)

This also will always be assumed to hold, and if no ambiguity arises x^a will be written, abusively, in place of x'^a.

Equations (3.4) and (3.5) being assumed, the map ψ is fixed completely if the coordinates y'^A_b are given as functions, say ψ^A_b, on $J^1(M,N_1) \times N_2$. Thus under ψ,

$$y'^A_b = \psi^A_b(x^a, z^\mu, z^\mu_a, y^B).$$ (3.6)

Now suppose that maps $f : M \to N_1$ and $g : M \to N_2$ are given, with local coordinate presentations

$$z^\mu = f^\mu(x), \qquad y^A = g^A(x) .$$

Then there are two different ways of constructing $\partial g^A/\partial x^b$, namely (1) by differentiating $g^A(x)$, and (2) by substituting into (3.6). There is no reason why for arbitrary ψ^A_b the two procedures should yield the same result – it is not in general the case that

$$\partial_b g^A(x) = \psi^A_b(x^a, f^\mu(x), \partial_a f^\mu(x), g^B(x)) ,$$ (3.7)

which, in coordinate-free terms, would be

$$j^1 g = \psi \circ (j^1 f \times g) \circ \Delta(M) .$$ (3.8)

In fact, given ψ, a map g satisfying (3.8) can exist only if the integrability conditions for it are satisfied. These integrability conditions, which follow from

$$\partial_c \, \partial_b \, g^A(x) - \partial_b \, \partial_c \, g^A(x) = 0,$$

are

$$g^A_{\ bc} = 0 \tag{3.9}$$

where

$$g^A_{\ bc} := \left[\frac{\partial \psi^A_{\ b}}{\partial x^c} + \frac{\partial \psi^A_{\ b}}{\partial z^\mu} \frac{\partial f^\mu}{\partial x^c} + \frac{\partial \psi^A_{\ b}}{\partial z^\mu_{\ d}} \frac{\partial^2 f^\mu}{\partial x^d \partial x^c} + \frac{\partial \psi^A_{\ b}}{\partial y^B} \psi^B_{\ c} \right]$$

$$\tag{3.10}$$

$$- \left[\frac{\partial \psi^A_{\ c}}{\partial x^b} + \frac{\partial \psi^A_{\ c}}{\partial z^\mu} \frac{\partial f^\mu}{\partial x^b} + \frac{\partial \psi^A_{\ c}}{\partial z^\mu_{\ d}} \frac{\partial^2 f^\mu}{\partial x^d \partial x^b} + \frac{\partial \psi^A_{\ c}}{\partial y^B} \psi^B_{\ b} \right].$$

Here $\psi^A_{\ b}$ and its derivatives are evaluated on the arguments exhibited in (3.7). These are conditions on the functions f^μ and their derivatives up to the second, in which the functions $g^A(x)$ appear, but not their derivatives, the latter having been eliminated with the help of (3.7). One would therefore expect to be able to write these conditions in coordinate-free form as equations on $J^2(M,N_1) \times N_2$, where exactly the right combinations of coordinates for the representation of these functions and derivatives are available.

We shall describe three different coordinate-free formulations of the integrability conditions, the first in terms of

exterior ideals of differential forms, the second in terms of
modules of 1-forms (Pfaffians), the last in terms of vector
fields. Each of these formulations has advantages in particular
applications. First we develop the necessary machinery,
generalizing from section 2.

First of all, the projection

$$\mathrm{pr}_1 \ : \ J^2(M,N_1) \times N_2 \to J^2(M,N_1)$$

induces on $J^2(M,N_1) \times N_2$ a naturally defined module $\mathrm{pr}_1{}^* \Omega^2(M,N_1)$.

Next, the projection

$$\pi_1^2 \ : \ J^2(M,N_1) \to J^1(M,N_1)$$

may be extended to

$$\widetilde{\pi}_1^2 \ := \ \pi_1^2 \times \mathrm{id}_{N_2} \ : \ J^2(M,N_1) \times N_2 \to J^1(M,N_1) \times N_2$$

and then

$$\psi \circ \widetilde{\pi}_1^2 \ : \ J^2(M,N_1) \times N_2 \to J^1(M,N_2)$$

induces on $J^2(M,N_1) \times N_2$ the module $\widetilde{\pi}_1^{2*} \ \psi^* \ \Omega^1(M,N_2)$. The sum of
these two induced modules is denoted $\widetilde{\Omega}^{2,\psi}$:

$$\widetilde{\Omega}^{2,\psi} \ := \ \mathrm{pr}_1{}^* \Omega^2(M,N_1) \ + \ \widetilde{\pi}_1^{2*} \psi^* \Omega^1(M,N_2). \tag{3.11}$$

As in section 2, let θ^μ and θ^μ_a denote standard contact
forms on $J^2(M,N_1)$ and let θ'^A denote standard contact forms on
$J^1(M,N_2)$. Where no ambiguity arises, we shall denote $\mathrm{pr}_1{}^* \theta^\mu$,

$\mathrm{pr_1}^*\theta^\mu{}_a$ and $\tilde{\pi}_1^{2*}\psi^*\theta'^A$ by θ^μ, $\theta^\mu{}_a$ and θ^A respectively. Then θ^μ and $\theta^\mu{}_a$ comprise a basis for $\mathrm{pr_1}^*\Omega^2(M,N_1)$, and θ^A comprise a basis for $\tilde{\pi}_1^{2*}\psi^*\Omega^1(M,N_2)$. Moreover, writing $\psi^A{}_b$ for $\tilde{\pi}_1^{2*}\psi^A{}_b$ here and below, the θ^A have the form

$$\theta^A = \mathrm{dy}^A - \psi^A{}_b \; \mathrm{dx}^b \; , \tag{3.12}$$

from $(3.4) - (3.6)$. Notice that any form which is a linear combination of the dx^a only is independent of the forms in (3.11).

The idea of the total derivative, and the associated operators $D^{(k)}{}_a$, may readily be generalized from $J^k(M,N_1)$ to $J^k(M,N_1) \times N_2$. All that is necessary is to include differentiation with respect to the coordinates on N_2. However, a coordinate-free formulation of the generalization of a prolongation requires a good deal of additional apparatus, because the generalized total derivatives do not commute, and we shall proceed from the formula (2.9) for the differential of a function, giving the generalized prolongation only in coordinates. Here the constructions are for the case $k = 2$ only; no more is needed in this section. The general case is dealt with in section 4.

Imitating the construction of $D^{(2)}{}_a$, let

$$\tilde{D}^{(2)}{}_a = D^{(2)}{}_a + \psi^B{}_a \; \frac{\partial}{\partial y^B}$$

$$= \frac{\partial}{\partial x^a} + z^\mu{}_a \; \frac{\partial}{\partial z^\mu} + z^\mu{}_{ad} \; \frac{\partial}{\partial z^\mu{}_d} + \psi^B{}_a \; \frac{\partial}{\partial y^B} \; . \tag{3.13}$$

Among the properties of this operator we note

1. If u is any function on $J^1(M, N_1) \times N_2$, then

$$d(\tilde{\pi}_1^{2*} u) = \tilde{D}^{(2)}_a (\tilde{\pi}_1^{2*} u) \, dx^a, \quad \text{mod } \tilde{\Omega}^{2, \psi} \quad . \tag{3.14}$$

This equation may be used as a definition of $\tilde{D}^{(2)}_a$, the local charts being given.

2. $$[\tilde{D}^{(2)}_a, \tilde{D}^{(2)}_b] = (\tilde{D}^{(2)}_a \psi^B_b - \tilde{D}^{(2)}_b \psi^B_a) \frac{\partial}{\partial y^B} \quad . \tag{3.15}$$

This formula plays a prominent part later on.

3. If X is a vector (field) of the form

$$X = X^a \, \tilde{D}^{(2)}_a$$

then $X \lrcorner \tilde{\Omega}^{2, \psi} = 0,$ \hfill (3.16)

but this is not enough to characterise $\tilde{D}^{(2)}_a$, because in addition

$$\frac{\partial}{\partial z^\mu_{ab}} \lrcorner \tilde{\Omega}^{2, \psi} = 0. \tag{3.17}$$

4. $$\tilde{D}^{(2)}_a \lrcorner d\theta^\mu = \theta^\mu_a , \tag{3.18}$$

$$\tilde{D}^{(2)}_a \lrcorner d\theta^\mu_b = dz^\mu_{ab} , \tag{3.19}$$

$$\tilde{D}^{(2)}_a \lrcorner d\theta^A = (\tilde{D}^2_b \psi^A_a - \tilde{D}^2_a \psi^A_b) \, dx^b \mod \tilde{\Omega}^{2, \psi}. \tag{3.20}$$

The distribution generated by the $\tilde{D}^{(2)}_a$ will be denoted $\tilde{\Delta}^{2, \psi}$ (Distributions or differential systems are defined in section 5). In virtue of (3.14) it is coordinate-independent.

The definition of a differential equation is easily generalized from $J^k(M,N)$ to $J^k(M,N_1) \times N_2$. A differential equation of order k is a submanifold $\tilde{Z}$ of $J^k(M,N_1) \times N_2$ which is the zero set of a finitely generated ideal $\tilde{\Sigma}_Z$ of functions on $J^k(M,N_1) \times N_2$. If $F_1, \ldots, F_q$ is a set of generators, then the system of equations may be written in terms of the coordinate presentations of these functions - say, in standard coordinates

$$F_\alpha(x^a, z^\mu, z^\mu_{\ b}, \ldots, z^\mu_{\ b_1 \ldots b_k}; y^A) = 0,$$

$$\alpha = 1, \ldots, q. \tag{3.21}$$

The generators define a map

$$F : J^k(M,N_1) \times N_2 \to \mathbb{R}^q$$

and a solution of the system is a pair of maps $f \in C^\infty(M,N_1)$, $g \in C^\infty(M,N_2)$ such that

$$F \circ (j^k f \times g) \circ \Delta(M) = 0. \tag{3.22}$$

Again we shall refer to the equations themselves, as well as to the zero set $\tilde{Z}$, as "the system of differential equations".

With this additional machinery available, we return to consideration of the integrability conditions (3.9). In the first place, comparing (3.10) with (3.13), one sees that (3.10) may be written

$$g^A_{\ bc} = (\tilde{D}^{(2)}_{\ c} \psi^A_{\ b} - \tilde{D}^{(2)}_{\ b} \psi^A_{\ c}) \circ (j^2 f \times g) \circ \Delta(M). \tag{3.23}$$

38

Consequently the pair (f,g) must be the solution of the system

$$\tilde{D}^{(2)}{}_c \psi^A{}_b - \tilde{D}^{(2)}{}_b \psi^A{}_c = 0 \tag{3.24}$$

on $J^2(M,N_1) \times N_2$. Now from (3.12),

$$d\theta^A = dx^b \wedge d\psi^A{}_b , \tag{3.25}$$

which by (3.14) may be written

$$d\theta^A = \tilde{D}^{(2)}{}_c \psi^A{}_b \; dx^b \wedge dx^c \bmod I(\tilde{\Omega}^2{}^{,\psi}) \tag{3.26}$$

(we continue the abuse of writing $\psi^A{}_b$ as an abbreviation for $\tilde{\pi}_1^{2*}\psi^A{}_b$). Comparing (3.26) with (3.24) and recalling that the θ^A generate $\tilde{\pi}_1^{2*}\psi^* \; \Omega^1(M,N_2)$ one may conclude that the system (3.24) is equivalent to

$$\tilde{\pi}_1^{2*}\psi^* d\Omega^1(M,N_2) \subset I(\tilde{\Omega}^2{}^{,\psi}) . \tag{3.27}$$

This is the coordinate-free formulation of the integrability conditions in terms of ideals of differential forms.

It is sometimes convenient to have these conditions formulated entirely in terms of the properties of Pfaffian modules, as follows:

Let

$$\hat{\Omega}^2{}^{,\psi} = \tilde{\Omega}^2{}^{,\psi} + \{X \; \lrcorner \; \tilde{\pi}_1^{2*}\psi^* d\Omega^1(M,N_2) \mid X \; \lrcorner \; \tilde{\Omega}^2{}^{,\psi} = 0\}. \tag{3.28}$$

Thus $\hat{\Omega}^2{}^{,\psi}$ is constructed by adding to $\tilde{\Omega}^2{}^{,\psi}$ the contractions of forms on the left hand side of (3.27) with vectors which annihilate $\widetilde{\Omega}^{2,\psi}$.

Then by (3.20), (3.27) is equivalent to

$$\hat{\Omega}^{2,\psi} \subset \widetilde{\Omega}^{2,\psi} . \tag{3.29}$$

Finally, from (3.13), (3.15) and (3.24) it is seen at once that the integrability conditions $\widetilde{Z}$ may be characterised as the submanifold on which the distribution $\widetilde{\Delta}^{2,\psi}$ is completely integrable.

In general, the integrability conditions are partial differential equations for the z's, with the y's entering as parameters. However, in many interesting cases, the y's appear only trivially, and the integrability conditions may be identified as a system of partial differential equations for the z's only. More precisely, it occurs in these cases that there exists a system of differential equations Z on $J^2(M,N_1)$ such that

$$Z \times N_2 \subset \widetilde{Z}. \tag{3.30}$$

Even when the integrability conditions for ψ are satisfied, the y's determined by ψ are not arbitrary. Any function g whose 1-jet lies in the image of ψ must satisfy (3.7) for some choice of f's. In the interesting cases, it turns out not only that one can eliminate the y's from the integrability conditions, obtaining equations for the f's alone, but also that one can eliminate the z's from the integrability conditions and their prolongations, obtaining equations for the g's alone. To formulate this in jet bundle language we need to generalize from section 2 the construction of a prolongation. As mentioned above, the coordinate-free formulation requires a good deal more machinery, so we forego it.

Again let

$$\psi : J^1(M,N_1) \times N_2 \to J^1(M,N_2)$$

be a map for which (3.2) and (3.3) commute, defined in standard coordinates by (3.4) - (3.6). A map

$$\psi^1 : J^2(M,N_1) \times N_2 \to J^2(M,N_2)$$

is said to be compatible with ψ if the diagram

$$
\begin{array}{ccc}
J^2(M,N_1) \times N_2 & \xrightarrow{\;\;\psi^1\;\;} & J^2(M,N_2) \\
\Big\downarrow{\tilde{\pi}^2_1} & & \Big\downarrow{\pi^2_1} \\
J^1(M,N_1) \times N_2 & \xrightarrow{\;\;\psi\;\;} & J^1(M,N_2)
\end{array}
\qquad (3.31)
$$

commutes. A map ψ^1 compatible with ψ is completely determined by the specification of functions ψ^A_{bc} on $J^2(M,N_1) \times N_2$ such that under ψ^1,

$$y'^A_{bc} = \psi^A_{bc}(x^a, z^\mu, z^\mu_a, z^\mu_{ad}, y^B). \qquad (3.32)$$

Here the y'^A_{bc} are standard coordinates on $J^2(M,N_2)$, and the arguments of ψ^A_{bc} are standard coordinates on $J^2(M,N_1) \times N_2$.

41

The appropriate choice of the functions $\psi^A_{\ bc}$ is

$$\psi^A_{\ bc} = \widetilde{D}^{(2)}_{\ (b}\ \psi^A_{\ c)} \quad , \tag{3.33}$$

where ( ) denotes symmetrization of the enclosed indices. The map ψ^1 defined by (3.31) – (3.33) will be called the first prolongation of ψ.

This completes the theoretical development. We proceed to consider some examples, and conclude this section with a tentative definition of Bäcklund maps and Bäcklund transformations.

Example 3.1. The sine–Gordon equation: Bäcklund's original map.

Let dim M = 2, dim N_1 = dim N_2 = 1. Let indices taking only the value 1 be omitted, except that $\pi^{2*}_1\psi^*\theta^{1'}$ is denoted θ^1, to distinguish it from θ^μ with μ = 1. The coordinates are x^1 and x^2 on M, z on N_1 and y = y' on N_2.

The map originally constructed by Bäcklund [4], [5] is (cf. (1.3))

$$y'_1 = \psi_1 \ := z_1 + 2a \sin \tfrac{1}{2}(y + z)$$

$$y'_2 = \psi_2 \ := -z_2 + 2a^{-1} \sin \tfrac{1}{2}(y - z) \tag{3.34}$$

where a is a non–zero parameter. The module $\widetilde{\Omega}^{2,\psi}$ is generated by

$$\theta = dz - z_1 dx^1 - z_2 dx^2$$

(as in Example 2.2),

$$\theta_1 = dz_1 - z_{11}dx^1 - z_{12}dx^2 \ ,$$

42

and $\theta_2 = dz_2 - z_{21}dx^1 - z_{22}dx^2$,

together with

$$\theta^1 = dy - (z_1 + 2a \sin \tfrac{1}{2}(y + z)) \, dx^1$$

$$- (- z_2 + 2a^{-1} \sin \tfrac{1}{2}(y - z)) \, dx^2 \ . \tag{3.35}$$

The operators $\tilde{D}^{(2)}_a$ have the explicit forms

$$\tilde{D}^{(2)}_1 = \frac{\partial}{\partial x^1} + z_1 \frac{\partial}{\partial z} + z_{11} \frac{\partial}{\partial z_1} + z_{12} \frac{\partial}{\partial z_2}$$

$$+ (z_1 + 2a \sin \tfrac{1}{2}(y + z)) \frac{\partial}{\partial y}$$

$$\tag{3.36}$$

$$\tilde{D}^{(2)}_2 = \frac{\partial}{\partial x^2} + z_2 \frac{\partial}{\partial z} + z_{21} \frac{\partial}{\partial z_1} + z_{22} \frac{\partial}{\partial z_2}$$

$$+ (- z_2 + 2a^{-1} \sin \tfrac{1}{2}(y - z)) \frac{\partial}{\partial y}$$

and the integrability condition for (3.34), obtained by substituting in (3.24), is the single equation

$$\left[\frac{\partial \psi_1}{\partial x^2} + z_2 \frac{\partial \psi_1}{\partial z} + z_{21} \frac{\partial \psi_1}{\partial z_1} + z_{22} \frac{\partial \psi_1}{\partial z_2} + (-z_2 + 2a^{-1} \sin\tfrac{1}{2}(y-z)) \frac{\partial \psi_1}{\partial y} \right]$$

$$- \left[\frac{\partial \psi_2}{\partial x^1} + z_1 \frac{\partial \psi_2}{\partial z} + z_{11} \frac{\partial \psi_2}{\partial z_1} + z_{12} \frac{\partial \psi_2}{\partial z_2} + (z_1 + 2a \sin\tfrac{1}{2}(y+z)) \frac{\partial \psi_2}{\partial y} \right] = 0$$

which yields the sine-Gordon equation (2.15), namely

$$z_{12} - \sin z = 0 \tag{3.37}$$

as integrability condition. Notice that it does not depend on y.

The first prolongation ψ^1 of ψ, obtained by substituting in
(3.33), is given by

$$y'_{11} = \psi_{11} = \widetilde{D}^{(2)}_{1} \psi_1 = z_{11} + 2a\ z_1 \cos\tfrac{1}{2}(y+z) + a^2 \sin(y+z),$$

$$y'_{12} = \psi_{12} = \tfrac{1}{2}(\widetilde{D}^{(2)}_{1} \psi_2 + \widetilde{D}^{(2)}_{2} \psi_1) = \sin y, \qquad\qquad (3.38)$$

$$y'_{22} = \psi_{22} = \widetilde{D}^{(2)}_{2} \psi_2 = -\,z_{22} - 2a^{-1} z_2 \cos\tfrac{1}{2}(y-z) + a^{-2}\sin(y-z).$$

Observe that ψ^1 is not surjective. Writing y' for y in the second of (3.38), one finds the equation

$$y'_{12} - \sin y' = 0 \quad .$$

Thus the image of ψ^1 is a differential equation Z' on $J^2(M,N_2)$ which does not depend on z. Moreover it is identical with the integrability condition, a situation which is described in the literature by calling the map (3.34) an "auto–Bäcklund transformation". $\qquad\qquad\qquad\qquad\qquad$ □

Example 3.2. The sine–Gordon equation: inverse scattering Riccati equation. The inverse scattering equations for the sine–Gordon equation [1], [2], written as a Riccati equation, yield another map with the sine–Gordon equation as integrability condition. With the same notation and conventions as in the preceding example, the map is

$$y'_1 = \tfrac{1}{2} a \sin z.(1 - y^2) + a \cos z.y,$$

$$\qquad\qquad\qquad\qquad\qquad\qquad\qquad\qquad (3.39)$$

$$y'_2 = a^{-1}y - \tfrac{1}{2} z_2(1 + y^2) \quad .$$

In order to construct ψ^1 it is convenient to change coordinates on N_2, writing

$$v := 2 \tan^{-1} y' \, .$$

Then on $J^1(M,N_2)$,

$$v_a = 2(1 + y'^2)^{-1} y'_a \quad .$$

If also (since $y = y'$)

$$\hat{\psi}_a := (1 + y^2)^{-1} \psi_a$$

the map may be rewritten

$$v_1 = \hat{\psi}_1 = a \sin(v + z) \, ,$$

$$\tag{3.40}$$

$$v_2 = \hat{\psi}_2 = a^{-1} \sin v - z_2 .$$

The first prolongation $\hat{\psi}^1$ of $\hat{\psi}$ is determined by $v_{ab} = \tilde{D}^{(2)}_{(a} \hat{\psi}_{b)}$, where $\tilde{D}^{(2)}_a = D^{(2)}_a + \hat{\psi}_a \frac{\partial}{\partial v}$. It is given by

$$v_{11} = \hat{\psi}_{11} = a \cos(v + z) . (a \sin(v + z) + z_1)$$

$$v_{12} = \hat{\psi}_{12} = \tfrac{1}{2}(\sin(2v + z) - z_{12}) \tag{3.41}$$

$$v_{22} = \hat{\psi}_{22} = \cos v \sin(v + z) - z_{22} \quad .$$

If $\hat{\psi}$ is restricted to the differential equation

$\tilde{Z} : z_{12} - \sin z = 0$ on $J^2(M,N_1) \times N_2$, then from the second of

(3.41), $v_{12} = \sin v \cos(v + z)$, while from the first of (3.40),

$v_1 = a \sin(v + z)$, so that elimination of $v + z$ yields [10]

$$v_{12} - (1 - a^{-2} v_1{}^2)^{\frac{1}{2}} \sin v = 0 \quad , \tag{3.42}$$

which shows that although $\hat{\psi}^1$ is surjective, the image of $\psi^1\big|_{\tilde{z}}$ is a differential equation which does not depend on z. It will be shown in section 7 that 2v is the difference of two solutions of the sine-Gordon equation.

Example 3.3. With dimensions and notation as in example 3.1. the Liouville transformation [52] is the map

$$y_1' = \psi_1 := z_1 + \beta \exp \tfrac{1}{2}(y + z)$$

$$y_2' = \psi_2 := - z_2 - 2\beta^{-1} \exp \tfrac{1}{2}(y - z) \tag{3.43}$$

where β is a non-zero parameter.

The integrability condition is found to be

$$z_{12} = 0 \quad , \tag{3.44}$$

and the first prolongation is

$$y_{11}' = \psi_{11} = z_{11} + \beta z_1 \exp \tfrac{1}{2}(y + z) + \tfrac{1}{2} \beta^2 \exp(y + z) \ ,$$

$$y_{12}' = \psi_{12} = - \exp y \ , \tag{3.45}$$

$$y_{22}' = \psi_{22} = - z_{22} + 2\beta^{-1} z_2 \exp \tfrac{1}{2}(y - z) + 2\beta^{-2} \exp(y - z).$$

Again the integrability condition is independent of y, and again the prolongation is not surjective; its image is the zero set of

$$y_{12}' + \exp y' = 0 \quad . \qquad\qquad \Box$$

46

Example 3.4. The "generalized Bäcklund-type transformations" of
Rogers [64] are defined by maps of the form (3.6), with (3.4) and
(3.5) also assumed to hold (equation (2.6) of [64]). In this case
dim M = 2, while dim N_1 = dim N_2 = n, with n > 1 in general.
Translated into our notation, the maps considered by Rogers are
linear in the z's and their derivatives and in the y's, with coefficients which may be functions of the x's. They include as a
special case the transformation found by Loewner [56] for the
reduction of the hodograph equations of gas dynamics to canonical
form. Rogers considers various other special cases in his paper.

Rogers's maps may be written, in our notation

$$y'^A_{\ b} = A^A_{\ \rho b}{}^c\, z^\rho_{\ c} + B^A_{\ \rho b}\, z^\rho + C^A_{\ Bb}\, y^B + D^A_{\ b} \qquad (3.46)$$

with the A's, B's, C's and D's functions lifted from M, that is,
functions of the x's (and without any particular assumption about
dim M). He then specializes the A's so that the integrability
conditions are actually first-order in the z's, by a choice of
A's which generalizes to the form

$$A^A_{\ \rho b}{}^c = A^A_{\ \rho}\, \delta^c_b \qquad ,$$

where δ^c_b is the Kronecker symbol. He specializes further, to
ensure that the integrability conditions do not contain the y's,
by a choice of C's which generalizes to

$$\partial_d\, C^A_{\ Bb} - \partial_b\, C^A_{\ Bd} + C^A_{\ Db}\, C^D_{\ Bd} - C^A_{\ Dd}\, C^D_{\ Bb} = 0. \qquad (3.47)$$

47

This implies that the C's must be of the form

$$C^A_{Bb} = G^C_B \, \partial_b \, g^A_C$$

where $[g^A_C]$ is any nowhere-singular matrix of functions lifted from M, and $[G^C_B]$ is its inverse. The integrability conditions for (3.46) are then

$$\alpha^B_{\rho[b} \, z^\rho_{a]} + \beta^B_{\rho[ab]} \, z^\rho + \zeta^B_{[ab]} = 0 \ , \tag{3.48}$$

where square brackets denote antisymmetrization of the enclosed indices, and the α's, β's and ζ's, which are all functions lifted from M, are given by

$$\alpha^B_{\rho b} = \partial_b (G^B_A \, A^A_\rho) - G^B_A \, B^A_{\rho b}$$

$$\beta^B_{\rho ab} = \partial_b (G^B_A \, B^A_{\rho a})$$

$$\zeta^B_{ab} = \partial_b (G^B_A \, D^A_a) \ .$$

Now suppose that the number of y's is the same as the number of z's. The matrix $[A^A_\rho]$ is then square, and may be assumed non-singular. Let $[a^\rho_A]$ denote its inverse. If the further conditions

$$\beta^B_{\sigma[ab]} = \alpha^B_{\sigma[b} \, a^\rho_{|A|} \, B^A_{\rho|a]} \tag{3.49}$$

are satisfied then from (3.48) one may deduce partial differential equations for the y's, namely

$$\alpha^B_{\rho[b} \, a^\rho_{|A|} (y^A_{a]} - C^A_{|B|a]} \, y^B - D^A_{a]}) + \zeta^B_{[ab]} = 0,$$

from which the z's have been eliminated. □

These examples suggest the following tentative definition: a map

$$\psi \; : \; J^1(M,N_1) \times N_2 \to J^1(M,N_2)$$

is a Bäcklund map if its integrability conditions comprise a system of differential equations $\tilde{Z}$ on $J^2(M,N_1) \times N_2$. If there is a system Z of differential equations on $J^2(M,N_1)$ such that $Z \times N_2 \subset \tilde{Z}$, as in (3.30), then ψ is called an ordinary Bäcklund map for Z. If the image of its first prolongation is (the zero set of) a system of differential equations Z' on $J^2(M,N_2)$, then the correspondence between Z and Z' is called the Bäcklund transformation determined by the ordinary Bäcklund map ψ. In example 3.1, (3.34), together with (3.4) and (3.5), defines an ordinary Bäcklund map with integrability condition Z: $z_{12} - \sin z = 0$ and image $Z' : y'_{12} - \sin y' = 0$; the transformation between the two is a Bäcklund transformation. In example 3.3, (3.43), together with (3.31) to (3.33), defines an ordinary Bäcklund map with integrability condition Z : $z_{12} = 0$ and image $Z' : y'_{12} + \exp y' = 0$; the transformation is again a Bäcklund transformation. In example 3.4, (3.46), together with (3.4) and (3.5) defines a Bäcklund map which is ordinary only if (3.48) is satisfied. Clairin [11] discusses Bäcklund maps which are not ordinary, so that $\tilde{Z}$ is not of the form (3.30), but all the Bäcklund maps which are known for the applications we are interested in are ordinary ones, so in this paper we confine ourselves to those.

However, the definition given above is not sufficiently general to embrace all known examples. A generalization of the procedure described above, in which both the integrability equation and the prolongation equation are of third order, has been invented recently by Wahlquist and Estabrook [72] to treat the Korteweg-de Vries equation. In the next section we generalize the above treatment in an obvious way to encompass this example and others.

Section 4

BACKLUND MAPS : GENERAL CASE

The generalization is to allow the first derivatives of the new
dependent variables to depend on derivatives of the old dependent
variables higher than the first. We proceed as far as possible
along the lines laid down in section 3:

Let M, N_1 and N_2 be C^∞ manifolds, and let

$$\psi : J^h(M,N_1) \times N_2 \to J^1(M,N_2) \tag{4.1}$$

be a C^∞ map. In the following, it is to be understood that h is
a fixed positive integer, while k, ℓ, m, n, q and s are non-
negative integers whose ranges are specified in context and are
not subject to the range and summation conventions. The development of section 3 may be recovered by setting $h = 1$. In the
applications, as before,

M is the space of independent variables,

N_1 is the space of the old dependent variables, and

N_2 is the space of the new dependent variables.

It is again assumed that ψ acts trivially on N_2, which is to
say that

$$\beta \circ \psi = pr_2 \qquad\qquad (4.2)$$

(cf. (3.2)), where, as before, β is the target projection.
It is assumed also that ψ acts trivially on M, which is to say
that

$$\alpha \circ \psi = \alpha \circ pr_1 \qquad\qquad (4.3)$$

(cf. (3.3)), where, as before, α is the source projection.
On the domain of ψ, choose local coordinates

x^a on M,   a, b, c, ... = 1, ..., dim M

z^μ on N_1, μ, ρ, ...   = 1, ..., dim N_1

y^A on N_2,   A, B, C, ... = 1, ..., dim N_2

(as in section 3), and (for each ℓ) standard association coordinates

$$z^\mu_a, \; z^\mu_{ab}, \; \ldots, \; z^\mu_{a_1 \ldots a_\ell} \; ,$$

pulled back from $J^\ell(M, N_1)$ by $pr_1{}^*$. The fact that
$z^\mu_a, \; \ldots, \; z^\mu_{a_1 \ldots a_k}$ are written for coordinates on $J^k(M, N_1) \times N_2$,
for $k \leqslant \ell$, as well as on $J^\ell(M, N_1) \times N_2$, does not lead to any
ambiguity.

On the codomain of ψ, choose local coordinates

x'^a on M

y'^A on N_2

and (for each ℓ) standard associated coordinates

$$y'^A_a, \; y'^A_{ab}, \; \ldots, \; y'^A_{a_1 \ldots a_\ell}$$

on $J^\ell(M, N_2)$.

In virtue of (4.2) and (4.3), choose the coordinates on M and N_2 so that

$$x'^a = x^a, \tag{4.4}$$

$$y'^A = y^A, \tag{4.5}$$

and write x^a in place of x'^a, y^A in place of y'^A, where it causes no ambiguity.

Equations (4.4) and (4.5) being assumed, the map ψ is fixed completely if the coordinates y'^A_b are given as functions, say ψ^A_b, on $J^h(M, N_1) \times N_2$. Thus under ψ,

$$y'^A_b = \psi^A_b(x^a, z^\mu, \ldots, z^\mu_{a_1 \ldots a_h}, y^B). \tag{4.6}$$

Exactly as in section 3, if maps $f : M \to N_1$ and $g : M \to N_2$ are given, with local coordinate presentations

$$z^\mu = f^\mu(x), \qquad y^A = g^A(x),$$

there is no reason why for arbitrary ψ^A_b the two procedures for construction of $\partial_b g^A$ should agree - it is not in general the case that

$$\partial_b g^A(x) = \psi^A_b(x^a, f^\mu(x), \ldots, \partial_{a_1 \ldots a_h} f^\mu(x), g^B(x)), \tag{4.7 (a)}$$

or, in coordinate-free terms, that

$$j^1 g = \psi \circ (j^h f \times g) \circ \Delta(M). \tag{4.7 (b)}$$

53

In fact, given ψ, a map g satisfying (4.7) (b) can exist only if the integrability conditions for (4.7) are satisfied. These integrability conditions are conveniently written in terms of a generalized total derivative operator.

Imitating the construction of $\tilde{D}^{(2)}_a$, let

$$\tilde{D}^{(\ell)}_a := D^{(\ell)}_a + \psi^B_a \frac{\partial}{\partial y^B} \tag{4.8}$$

$$= \frac{\partial}{\partial x^a} + z^\mu_a \frac{\partial}{\partial z^\mu} + \ldots + z^\mu_{aa_1 \ldots a_{\ell-1}} \frac{\partial}{\partial z^\mu_{a_1 \ldots a_{\ell-1}}} + \psi^B_a \frac{\partial}{\partial y^B} \, .$$

Then the integrability conditions are

$$g^A_{bc} = 0 \tag{4.9}$$

where

$$g^A_{bc} := \tilde{D}^{(h+1)}_c \psi^A_b - \tilde{D}^{(h+1)}_b \psi^A_c \tag{4.10}$$

and ψ^A_b and its derivatives are evaluated on the arguments exhibited in (4.7) (a), with the addition of $\partial_{a_1 \ldots a_{h+1}} f^\mu$.

The argument continues as in section 3, but with $J^1(M,N_1)$ everywhere replaced by $J^h(M,N_1)$. The definition of a system of differential equations given in section 3 needs no modification. Thus, for every ℓ, the projection

$$pr_1 : J^\ell(M,N_1) \times N_2 \to J^\ell(M,N_1)$$

induces on $J^\ell(M,N_1) \times N_2$ a naturally defined module $pr_1^* \Omega^\ell(M,N_1)$.

Then, for every non-negative ℓ and for $\ell \geqslant k \geqslant 0$, the projection π_k^ℓ may be extended to

$$\tilde{\pi}_k^\ell := \pi_k^\ell \times id_{N_2} \; : \; J^\ell(M,N_1) \times N_2 \to J^k(M,N_1) \times N_2,$$

whence it follows that for $m \geq \ell \geq k$,

$$\tilde{\pi}_k^\ell \circ \tilde{\pi}_\ell^m = \tilde{\pi}_k^m \; .$$

Moreover, for $\ell \geq h$,

$$\psi \circ \tilde{\pi}_h^\ell \; : \; J^\ell(M,N_1) \times N_2 \to J^1(M,N_2)$$

induces on $J^\ell(M,N_1) \times N_2$ the module $\tilde{\pi}_h^{\ell *}\psi^*\Omega^1(M,N_2)$. The sum of
the two induced modules on $J^\ell(M,N_1) \times N_2$, for $\ell \geq h$, is denoted
$\tilde{\Omega}^{\ell,\psi}$:

$$\tilde{\Omega}^{\ell,\psi} = pr_1{}^*\Omega^\ell(M,N_1) + \tilde{\pi}_h^{\ell *}\psi^*\Omega^1(M,N_2). \tag{4.11}$$

As in section 2, let the 1-forms defined in (2.7) (with k
replaced by ℓ) be standard contact forms on $J^\ell(M,N_1)$. As in
section 3, let θ'^A denote standard contact forms on $J^1(M,N_2)$.
Where no ambiguity arises, we shall denote $pr_1{}^*\theta^\mu{}_{a_1 \ldots a_k}$ by
$\theta^\mu{}_{a_1 \ldots a_k}$ and $\tilde{\pi}_h^{\ell *}\psi^*\theta'^A$ by θ^A. Then $\theta^\mu, \ldots, \theta^\mu{}_{a_1 \ldots a_{\ell-1}}$ comprise
a basis for $pr_1{}^*\Omega^\ell(M,N_1)$ and θ^A comprise a basis for
$\tilde{\pi}_h^{\ell *}\psi^*\Omega^1(M,N_2)$. Moreover, writing $\psi^A{}_b$ for $\tilde{\pi}_h^{\ell *}\psi^A{}_b$, here and sometimes below, the θ^A have the form

$$\theta^A = dy^A - \psi^A{}_b \, dx^b \; , \tag{4.12}$$

exactly as in (3.12). Again any form which is a linear

combination of the dx^a only is independent of the forms in (4.11). Notice that since $\pi_k^{\ell *} \widetilde{\Omega}^k(M,N_1) \subset \Omega^\ell(M,N_1)$ for $\ell \geqslant k$, $\widetilde{\pi}_k^{\ell *} \widetilde{\Omega}^{k,\psi} \subset \widetilde{\Omega}^{\ell,\psi}$ for $\ell \geqslant k \geqslant h$, so that $\widetilde{\Omega}^{k,\psi}$ may be regarded as a submodule of $\widetilde{\Omega}^{\ell,\psi}$.

We note the following properties of the operators $\widetilde{D}^{(\ell)}_a$:

1. If u is any function on $J^{\ell-1}(M,N_1) \times N_2$, then

$$d(\pi_{\ell-1}^{\ell *} u) = \widetilde{D}^{(\ell)}_a \, (\pi_{\ell-1}^{\ell *} u) \, dx^a, \quad \mod \widetilde{\Omega}^{\ell,\psi} \, . \tag{4.13}$$

This equation may be used as a definition of $\widetilde{D}^{(\ell)}_a$, the local charts being given.

2. $[\widetilde{D}^{(\ell)}_a, \widetilde{D}^{(\ell)}_b] = (\widetilde{D}^{(\ell)}_a \, \widetilde{\pi}_h^{\ell *} \psi^B_b - \widetilde{D}^{(\ell)}_b \, \widetilde{\pi}_h^{\ell *} \psi^B_a) \, \dfrac{\partial}{\partial y^B}$ for $\ell > h$.

$$\tag{4.14}$$

3. If X is a vector (field) of the form

$$X = X^a \, \widetilde{D}^{(\ell)}_a$$

then $X \lrcorner \widetilde{\Omega}^{\ell,\psi} = 0$, $\tag{4.15}$

but this is not enough to characterise $\widetilde{D}^{(\ell)}_a$, because in addition

$$\dfrac{\partial}{\partial z^\mu_{a_1 \ldots a_\ell}} \lrcorner \widetilde{\Omega}^{\ell,\psi} = 0 \, . \tag{4.16}$$

4. $\widetilde{D}^{(\ell)}_a \lrcorner \, d\theta^\mu = \theta^\mu_a,$

$\widetilde{D}^{(\ell)}_a \lrcorner \, d\theta^\mu_{a_1} = \theta^\mu_{aa_1} \, ,$

$\vdots$

$$\tag{4.17}$$

$\widetilde{D}^{(\ell)}_a \lrcorner \, d\theta^\mu_{a_1 \ldots a_{\ell-2}} = \theta^\mu_{aa_1 \ldots a_{\ell-2}} \, ,$

$$\widetilde{D}^{(\ell)}_a \ \lrcorner \ d\theta^{\mu}_{a_1 \ldots a_{\ell-1}} = dz^{\mu}_{a a_1 \ldots a_{\ell-1}} \ , \tag{4.18}$$

$$\widetilde{D}^{(\ell)}_a \ \lrcorner \ d\theta^A = (\widetilde{D}^{(\ell)}_b \ \psi^A_a - \widetilde{D}^{(\ell)}_a \ \psi^A_b) \ dx^b \ \mathrm{mod} \ \widetilde{\Omega}^{h+1,\psi}, \tag{4.19}$$

all for $\ell > h$.

The distribution generated by the $\widetilde{D}^{(\ell)}_a$ will be denoted $\Delta^{\ell,\psi}$.
In virtue of (4.13) it is coordinate-independent.

We return now to consideration of the integrability conditions
(4.10). It follows from (4.10) that the pair (f,g) must be a
solution of the system

$$\widetilde{D}^{(h+1)}_c \ \psi^A_b - \widetilde{D}^{(h+1)}_b \ \psi^A_c = 0 \ . \tag{4.20}$$

These are the integrability conditions, as differential equations
on $J^{h+1}(M,N_1) \times N_2$. As is apparent from (4.8), they are quasi-
linear, that is, linear in the $z^{\mu}_{b_1 \ldots b_{h+1}}$. This restriction to
quasi-linearity may be surmounted if the trivial action of ψ on
M, represented in (4.3), is dropped, but it is known that not all
systems of partial differential equations may be expressed as
integrability conditions ([26] section 299 for $h = 1$, dim $N_1 = 1$).
Now from (4.12), on $J^{h+1}(M,N_1) \times N_2$,

$$d\theta^A = dx^b \wedge d\psi^A_b \ , \tag{4.21}$$

which by (4.13) may be written

$$d\theta^A = \widetilde{D}^{(h+1)}_c \ \psi^A_b \, dx^b \wedge dx^c \ \mathrm{mod} \ I(\widetilde{\Omega}^{h+1,\psi}). \tag{4.22}$$

Comparing (4.22) with (4.20) and recalling that the θ^A

generate $\tilde{\pi}_h^{h+1*}\psi^*\Omega^1(M,N_2)$ one may conclude that the system (4.20) is equivalent to

$$\tilde{\pi}_h^{h+1*}\psi^*d\Omega^1(M,N_2) \subset I(\tilde{\Omega}^{h+1,\psi}) . \qquad (4.23)$$

This is the coordinate-free form of the integrability conditions in terms of ideals of differential forms. For the Pfaffian module formulation, let

$$\hat{\Omega}^{\ell,\psi} = \tilde{\Omega}^{\ell,\psi} + \{X \lrcorner \tilde{\pi}_h^{\ell*}\psi^*d\Omega^1(M,N_2) \mid X \lrcorner \tilde{\Omega}^{\ell,\psi} = 0\} \qquad (4.24)$$

for $\ell > h$.

Then by (4.19), (4.23) is equivalent to

$$\hat{\Omega}^{h+1,\psi} \subset \tilde{\Omega}^{h+1,\psi} . \qquad (4.25)$$

Finally, from (4.8), (4.14) and (4.20) it is seen that the integrability conditions $\tilde{Z}$ may be characterised as the submanifold on which the distribution $\Delta^{h+1,\psi}$ is completely integrable.

Again, in many interesting cases the y's appear only trivially, in the sense that there exists a system of differential equations Z on $J^{h+1}(M,N_1)$ such that

$$Z \times N_2 \subset \tilde{Z} . \qquad (4.26)$$

On the other hand, it may be possible to eliminate the z's, obtaining equations for the y's alone. To this end, successive prolongations of ψ are needed. Suppose that

$$\psi : J^h(M,N_1) \times N_2 \to J^1(M,N_2)$$

is a map satisfying (4.2) and (4.3), and defined in local

coordinates by (4.4) – (4.6). A map

$$\psi^1 : J^{h+1}(M,N_1) \times N_2 \to J^2(M,N_2)$$

is said to be compatible with ψ if the diagram

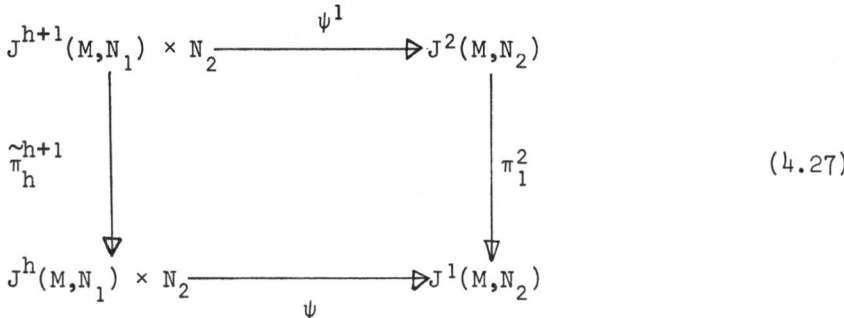

$$(4.27)$$

commutes. A map ψ^1 compatible with ψ is completely determined

by the specification of functions ψ^A_{bc} on $J^{h+1}(M,N_1) \times N_2$ such

that under ψ^1,

$$y'^A_{bc} = \psi^A_{bc} (x^a, z^\mu, \ldots, z^\mu_{a_1 \ldots a_{h+1}}, y^B).$$ (4.28)

The appropriate choice of the functions ψ^A_{bc} is

$$\psi^A_{bc} = \tilde{D}^{(h+1)}_{(b} \psi^A_{c)} ,$$ (4.29)

and the map ψ^1 defined by (4.27) – (4.29) will be called the

first prolongation of ψ. For s = 2, 3, ... the sth prolongation

of ψ is the map

$$\psi^s : J^{h+s}(M,N_1) \times N_2 \to J^{s+1}(M,N_2)$$ (4.30)

which is compatible with ψ^{s-1} via natural projections:

$$\pi^{s+1}_{s} \circ \psi^{s} = \psi^{s-1} \circ \tilde{\pi}^{h+s}_{h+s-1} \quad , \tag{4.31}$$

and which consequently determines ψ^{s} completely by the specifi-
cation of $\psi^{A}_{b_1 \ldots b_{s+1}}$ on $J^{h+s}(M,N_1) \times N_2$ such that under ψ^{s},

$$y'^{A}_{b_1 \ldots b_{s+1}} = \psi^{A}_{b_1 \ldots b_{s+1}} (x^{a}, z^{\mu}, \ldots, z^{\mu}_{a_1 \ldots a_{h+s}}, y^{B}).$$

$$\tag{4.32}$$

The appropriate choice of the functions $\psi^{A}_{b_1 \ldots b_{s+1}}$ is

$$\psi^{A}_{b_1 \ldots b_{s+1}} = \tilde{D}^{(h+s)}_{(b_1 b_2 \ldots b_s} \psi^{A}_{b_{s+1})} \quad . \tag{4.33}$$

We shall have occasion to make use of prolongations of the
integrability conditions also. The sth prolongation $\tilde{Z}^{s}$ of the
integrability conditions for ψ comprises the equations

$$\tilde{D}^{(h+s)}_{a_1 \ldots a_{r-2}[a_{r-1}} \psi^{A}_{a_r]} = 0 \tag{4.34}$$

with r taking all of the values 2, 3, ..., s + 1 successively
(r = 2 yields the integrability conditions themselves).

This completes the theoretical development for this section.
We now make a tentative generalization of the definition of a
Bäcklund map given in section 3.

A map

$$\psi : J^{h}(M,N_1) \times N_2 \rightarrow J^{1}(M,N_2)$$

is a Bäcklund map if its integrability conditions comprise a system of differential equations $\widetilde{Z}$ on $J^{h+1}(M,N_1) \times N_2$. If there is a system of differential equations Z on $J^{h+1}(M,N_1)$ such that

$$Z \times N_2 \subset \widetilde{Z},$$

then ψ is called an ordinary Bäcklund map for Z.

In some applications it occurs that one or more of the new independent variables are given functions of the old independent variables and their derivatives, say in the form

$$y^A = \chi^A(x^a, \ z^\mu, \ z^\mu_a, \ldots, \ z^\mu_{a_1 \ldots a_{h-1}}) \tag{4.35}$$

for some A's, from which is deduced

$$y^A_b = D^{(h)}_b \ \chi^A(x^a, \ z^\mu, \ldots, \ z^\mu_{a_1 \ldots a_h}) \ .$$

A map defined in this way naturally yields no integrability conditions, and it may be convenient to impose additional differential equations on $J^{h+1}(M,N_1)$ to make up for this. Therefore we extend the definition by allowing that if $\widetilde{T}$ is a submanifold of $J^h(M,N_1) \times N_2$ and ψ is a Bäcklund map with integrability conditions $\widetilde{Z}$ such that $\widetilde{T} \cap \widetilde{Z}$ is a submanifold of $\widetilde{Z}$ then $\psi\big|_{\widetilde{T}}$ is a Bäcklund map.

If there is a least integer s such that the image of $\psi^s\big|_{\widetilde{Z}^s}$ is a system of differential equations Z' on $J^{s+1}(M,N_2)$ then the correspondence between Z and Z' will be called the Bäcklund transformation determined by the Bäcklund map ψ. "Auto–Bäcklund" transformations will be characterized in section 6, where the appropriate machinery is developed.

If for some $h \geq 0$ and some M and N_1, there is given a system of partial differential equations Z on $J^{h+1}(M,N_1)$, then the Bäcklund problem for Z shall mean the determination of manifolds N_2 and maps $\psi : J^h(M,N_1) \times N_2 \to J^1(M,N_2)$ which are ordinary Bäcklund maps for Z.

We conclude this section with two examples.

Example 4.1. The KdV potential equation. The differential equation generated by the single function

$$F := z_{111} + z_2 - 6z_1{}^2 \qquad\qquad (4.36)$$

is called the potential equation for the Korteweg-deVries equation, because its first prolongation is generated by (4.36) and

$$
\begin{aligned}
F_1 &= z_{1111} + z_{12} - 12z_1 \, z_{11} \qquad &\text{(a)} \\
&&&(4.37) \\
F_2 &= z_{1112} + z_{22} - 12z_1 \, z_{12} \qquad &\text{(b)}
\end{aligned}
$$

and (4.37) (a) is just the Korteweg-deVries equation for $-z_1$.

A one-parameter family of Bäcklund maps whose integrability condition is the Korteweg-deVries potential equation (4.36) was discovered by Wahlquist and Estabrook [71]. Let dim $M = 2$, dim N_1 = dim N_2 = 1. Let indices taking only the value 1 be omitted, except that $\tilde{\pi}_2^{3*}\psi^*\theta'^1$ is denoted θ^1, to distinguish it from θ^μ with $\mu = 1$, which is denoted simply θ. The coordinates are x^1 and x^2 on M, z on N_1 and $y = y^1$ on N_2. The map discovered by Wahlquist and Estabrook is (with $h = 2$)

$$\psi(t) : J^2(M,N_1) \times N_2 \rightarrow J^1(M,N_2)$$

by

$$y_1' = \psi_1(t) = -z_1 - 2t + (y - z)^2$$

$$y_2' = \psi_2(t) = -z_2 + 4\{4t^2 + 2tz_1 - 2t(y-z)^2$$

$$+ z_1^2 + z_1(y-z)^2 + z_{11}(y-z)\}.$$

(4.38)

Here, for later convenience (Example 6.2), the parameter denoted k^2 by Wahlquist and Estabrook is written $2t$.

The integrability condition, formed by substitution into (4.20), is

$$(y - z)(z_{111} + z_2 - 6z_1^2) = 0 .$$

(4.39)

Since $y = z$ yields only the identity transformation, it is rejected, leaving the potential equation as integrability condition Z and showing that (4.38) is an ordinary Bäcklund transformation for it. The first prolongation is

$$y'_{11} = \psi_{11} = -z_{11} + 2(y - z)\{-2z_1 - 2t + (y - z)^2\}$$

$$y'_{12} = \psi_{12} = -z_{12} + 2(y - z)(16t^2 + 8t^2 z_1 - 2z_1^2 + z_{111} - z_2)$$

$$+ 8(y - z)^2 z_{11} + 8(y - z)^3 (z_1 - 2t)$$

63

$$y'_{22} = \psi_{22} = -z_{22} + 8tz_{12} + 2z_1 z_{12} + z_{112}(y - z) + z_{12}(y - z)^2$$

$$+ \{z_{11} + (2z_1 - 4t)(y-z)\} \times \qquad (4.40)$$

$$\times \{-2z_1 + 4(4t^2 + 2tz_1 - 2t(y-z)^2 + z_1^2$$

$$+ z_1(y-z)^2 + z_{11}(y-z))\}.$$

Prolonging once more, one finds after a short calculation that

$$\psi_{111} + \psi_2 - 6\psi_1^2 = -(z_{111} + z_2 - 6z_1^2) , \qquad (4.41)$$

so that the image of $\psi^2 \Big|_{Z^1 \times N_2}$ is the equation Z' defined by

$$y'_{111} + y'_2 - 6y'_1^2 = 0.$$

Thus the correspondence between Z and Z' is an auto-Bäcklund transformation. ☐

Example 4.2. A Bäcklund map whose integrability condition is the Korteweg-deVries equation

$$z_{111} + z_2 + 12zz_1 = 0 \qquad (4.42)$$

itself was discovered by Wahlquist and Estabrook [72]. It is

$$\psi : J^2(M,N_1) \times N_2 \to J^1(M,N_2)$$

by

$$y'_1 = \psi_1 := -2z - y^2$$

$$\text{(4.43)}$$

$$y'_2 = \psi_2 := 8z^2 + 4zy^2 + 2z_{11} - 4z_1 y.$$

The image under ψ^2 of the subset of $J^1(M, N_1) \times N_2$ defined by (4.42) is

$$y_{111} + y_2 - 6y^2 y_1 = 0, \tag{4.44}$$

which is called the Modified Korteweg-deVries equation (cf.(1.5)). □

Section 5

CONNECTIONS

To the reader familiar with the theory of connections, it may not
come as a surprise that some of the ideas which have been introduced may be formulated in the language of that theory. Equations
(3.15) for example, suggest that a connection is lurking somewhere.
In this section we propose two related formulations of Bäcklund
maps in terms of connections. In section 7 we employ these
formulations to give a geometric interpretation of the "KdV Lie
algebra" discovered by Wahlquist and Estabrook [72]. Similar
interpretations, not employing jet bundles, have been suggested
by R. Hermann. Our formulations are not as straightforward as we
should like, and we suspect that there must be another, more
direct, approach to Bäcklund maps in terms of connections yet to
be uncovered.

We begin by reproducing Ehresmann's definition of a connection, more or less as expounded by R. Hermann ([42] Ch. 4). This
version is convenient because in the applications it is natural
to introduce the connection first and the structure group after-
wards, so that one cannot begin, as is usual in most standard
treatments, by defining the connection on a principal bundle.

66

Let E and M be (smooth) manifolds, and $\pi : E \to M$ a surjective map of maximal rank. We shall give conditions for $E = (E,M,\pi)$ to be a local product fibre space, and define a connection to be an horizontal distribution on E. In the applications, M is the space of independent variables (space-time), as in the preceding sections, and E is either $M \times N_2$ or the product $J(M,N_1) \times N_2$, where $J(M,N_1)$ is the jet bundle of infinite order defined below. For the present, however, E is assumed to be finite-dimensional.

For every $\eta \in E$, assume that $\pi(\eta)$ has a neighbourhood U such that there exists a diffeomorphism

$$P_U : U \times F \to \pi^{-1}(U) \tag{5.1}$$

where F is a fixed manifold. Then $E = (E, M, \pi)$ is called a local product fibre space with typical fibre F. For every $x \in M$, the fibre $\pi^{-1}(x)$ is diffeomorphic to F.

A vector X tangent to E at η is called vertical if it is tangent to the fibre through η, that is to say, if $\pi_* X = 0$. The subspace of vertical vectors at η is denoted $V_\eta E$. A vector field is called vertical if it is vertical at each point.

If x^a are local coordinates on M and y^A are local coordinates on F, then $\pi^* x^a$ and $P_U^{-1*} y^A$, which may be chosen as local coordinates on E, will be abbreviated to x^a and y^A and called adapted coordinates. It follows that $\pi_*(\partial/\partial x^a) = \partial/\partial x^a$ and $\pi_*(\partial/\partial y^A) = 0$, so that every vertical vector field has the coordinate presentation $\xi^A \partial/\partial y^A$, where ξ^A are functions on E.

A distribution D on E is an assignment of a subspace $D_\eta E \subset T_\eta E$ at each $\eta \in E$. We shall be concerned only with regular distributions, for which dim $D_\eta E$ has a fixed value, independent of η, denoted dim D. Moreover we shall suppose that all the distributions which arise are differentiable, which means that each $\eta \in E$ has a neighbourhood W in which there are dim D differentiable vector fields spanning $D_\zeta E$ for each $\zeta \in W$. If X is a vector field on E and $X(\eta) \in D_\eta E$ for each η then one says that X belongs to D. A distribution D is called involutive if [X,Y] belongs to D whenever X and Y belong to D. If $i : E' \to E$ is the natural injection of a submanifold E' of E, and if $i*(T_\eta, E') = D_{i(\eta')} E$ for every $\eta' \in E'$, then E' is called an integral manifold of D. A distribution D on E is called integrable if an integral manifold of D passes through every point of E. Frobenius's theorem asserts that a distribution is integrable if and only if it is involutive [6].

An horizontal distribution H on E is a distribution which is complementary to the vertical in the sense that for each $\eta \in E$,

$$T_\eta E = V_\eta E \oplus H_\eta E .$$

It follows that

$$\dim H_\eta E = \dim M$$

and that

$H_\eta E \cap V_\eta E = 0$.

If H is a given horizontal distribution on E then any vector X tangent to E at any point η has a unique decomposition

$$X = hX + vX \tag{5.2}$$

where $hX \in H_\eta E$ and $vX \in V_\eta E$.

A vector X at η is called horizontal if it lies in $H_\eta E$. In adapted local coordinates, horizontal vectors have a basis of the form

$$H_a = \partial/\partial x^a + \Gamma_a^{\ A} \, \partial/\partial y^A \tag{5.3}$$

where $\Gamma_a^{\ A}$ are functions on E which determine, and are determined by, H. In these coordinates, the decomposition (5.2) of $X = \xi^a \, \partial/\partial x^a + \zeta^A \, \partial/\partial y^A$ is given by

$$X = \xi^a H_a + (\zeta^A - \xi^a \Gamma_a^{\ A}) \, \partial/\partial y^A \ . \tag{5.4}$$

A vector field X is called horizontal if it belongs to H. Thus an horizontal vector field is of the form

$$X = \xi^a H_a \tag{5.5}$$

where ξ^a are functions on E. A curve is called horizontal if its tangent vector at each point is horizontal.

The 1-forms which annihilate, and are annihilated by, horizontal vectors are called vertical forms. The vertical forms comprise a $C^\infty(E)$-module

69

$$H^* = \{\theta \in \Lambda^1(E) \mid X \lrcorner \theta = 0 \text{ for all } X \in H\} \,, \qquad (5.6)$$

which has a local basis in adapted coordinates of the form

$$\theta^A = dy^A - \Gamma_a{}^A \, dx^a \qquad (5.7)$$

where $\Gamma_a{}^A$ are the functions introduced in (5.3).

The $C^\infty(E)$-bilinear map

$$\Omega : T_\eta E \times T_\eta E \to V_\eta E$$

by $\quad (X_1, X_2) \mapsto v[hX_1, hX_2] \qquad (5.8)$

is called the curvature of H at η. The bilinearity may be verified by the computation

$$v[hX_1, h(fX_2)] = v(f[hX_1, hX_2] + (hX_1)f.(hX_2))$$

$$= fv[hX_1, hX_2] \,,$$

where f is any function on E.

It follows from Frobenius's theorem that an horizontal distribution is integrable if and only if its curvature vanishes.

From (5.8),

$$\Omega(H_a, H_b) = v[H_a, H_b] = v[\partial/\partial x^a + \Gamma_a{}^A \partial/\partial y^A, \partial/\partial x^b + \Gamma_b{}^B \partial/\partial y^B]$$

$$= \Omega_{ab}{}^B \partial/\partial y^B \qquad (5.9)$$

where

$$\Omega_{ab}{}^B := (\partial/\partial x^a + \Gamma_a{}^A \partial/\partial y^A)\Gamma_b{}^B - (\partial/\partial x^b + \Gamma_b{}^A \partial/\partial y^A)\, \Gamma_a{}^B \qquad (5.10)$$

70

are called the components of the curvature tensor.

If X is tangent to M at x, the horizontal lift of X to any point $\eta \in \pi^{-1}(x)$ is the unique horizontal vector $\bar{X} \in H_\eta E$ such that

$$\pi_* \bar{X} = X.$$

If in local coordinates

$$X = X^a(x) \; \partial/\partial x^a$$

then from (5.3) (writing $X^a(x)$ again for $X^a \circ \pi(\eta)$)

$$\bar{X} = X^a(x)(\partial/\partial x^a + \Gamma_a^A \; \partial/\partial y^A). \qquad (5.11)$$

If γ is a curve through $x \in M$, the horizontal lift of γ through $\eta \in \pi^{-1}(x)$ is the unique horizontal curve $\bar{\gamma}$ through η such that

$$\pi \circ \bar{\gamma} = \gamma \; ,$$

if it exists; η is said to be parallelly transported along γ to other points of $\bar{\gamma}$, which are sometimes called parallel translates of η.

If every curve on M has an horizontal lift through each point above it, the horizontal distribution is called a connection on E. From now on, we assume that all the horizontal distributions introduced are connections.

The functions Γ_a^A introduced in (5.3) are in this case called the connection coefficients. If the presentation of γ in local coordinates is $t \mapsto \gamma^a(t)$, and that of $\bar{\gamma}$ is $t \mapsto (\bar{\gamma}^a(t), \gamma^A(t))$, then from (5.11) it follows that

$$\bar{\gamma}^a(t) = \gamma^a(t) \tag{5.12}$$

and $\dfrac{d\gamma^A}{dt} = \Gamma_a^A(\gamma^b(t), \gamma^B(t)) \dfrac{d\gamma^a}{dt}$; $\tag{5.13}$

the latter are called the equations of parallel transport. If

$$\sigma : U \to E$$

is a (local) section of π, U an open set in M, then the covariant derivative $\nabla_X \sigma$ of σ with respect to a vector X tangent to U at x is defined by

$$\nabla_X \sigma = v(\sigma_* X) \ . \tag{5.14}$$

In local coordinates, if $X = X^a \ \partial/\partial x^a$ and

$$\sigma(x) = (x^a, \ \sigma^A(x)) \tag{5.15}$$

then

$$\nabla_X \sigma = (\nabla_X \sigma)^A \ \partial/\partial y^A$$

where

$$(\nabla_X \sigma)^A = X^a \ (\frac{\partial \sigma^A}{\partial x^a} - \Gamma_a^A(x, \ \sigma^B(x)) \ . \tag{5.16}$$

The section σ is called integrable if $\nabla_X \sigma = 0$ for all vector fields X tangent to U. It is no more than a restatement of Frobenius's theorem to remark that there is an integrable (local)

72

section through each point if and only if the curvature vanishes.

In the cases of interest, where the connection will be associated with a Bäcklund map, it may be possible to endow the local product fibre space with some additional structure:

(1) E may be a (differentiable) fibre bundle, with (Lie) structure group G acting on the fibre F; in this case the Lie algebra of G will be denoted $\underline{G}$ and the Lie algebra of vector fields on F which generates the action of G (and which is an homomorphic image of $\underline{G}$) will be denoted $\underline{G}_F$.

(2) the fibre bundle E may be a vector bundle, with F a vector space, G a subgroup of the group of linear transformations GL(F), and H a linear connection.

If E is a fibre bundle, the action of G must be compatible with the diffeomorphisms (5.1), which means that if

$$P_U : U \times F \to \pi^{-1}(U).$$

$$P'_U : U' \times F \to \pi^{-1}(U')$$

are two such diffeomorphisms, with U ∩ U' not empty, and if for each x ∈ U ∩ U',

$$P_x : F \to \pi^{-1}(x) \quad \text{by} \quad y \mapsto P_U(x,y)$$

$$P'_x : F \to \pi^{-1}(x) \quad \text{by} \quad y \mapsto P'_U(x,y)$$

then $P_x^{-1} \circ P_x' : F \to F$

73

is an action of some $g_x \in G$ on F, depending smoothly on x.

The possibility of endowing E with the structure of a fibre bundle is of interest here only if the action of G is compatible also with parallel transport by the connection H. This means that parallel transport, suitably composed with the diffeomorphisms (5.1), should yield an action of G. Explicitly: let x_1 and x_2 be points of M, γ a curve joining them, with $\gamma(t_1) = x_1$ and $\gamma(t_2) = x_2$. For each $\eta \in \pi^{-1}(x_1)$, let $\bar{\gamma}_\eta$ be the horizontal lift of γ through η. Then parallel transport along γ is given by

$$\tau_\gamma : \pi^{-1}(x_1) \to \pi^{-1}(x_2) \tag{5.17}$$

by $\quad \eta \mapsto \bar{\gamma}_\eta(t_2)$.

Let U_1 and U_2 be neighbourhoods of x_1 and x_2 respectively for which diffeomorphisms (5.1) are defined and let P_1 and P_2 be the maps $P_i : F \to \pi^{-1}(x_i)$ by $y \mapsto P_{U_i}(x_i, y)$. Then compatibility implies that

$$P_2^{-1} \circ \tau_\gamma \circ P_1 : F \to F \tag{5.18}$$

is an action of $g \in G$ on F. If this is the case, H is sometimes called a G-connection.

If H is a G-connection, then (5.18) imposes conditions on the connection coefficients, for it implies that to each curve γ through $x \in M$ there is a curve g_γ in G such that

$$\bar{\gamma}(t) = P_U(\gamma(t), g_\gamma(t)y) \tag{5.19}$$

where $g_\gamma(0) = e$, the identity of G, $\gamma(0) = x$, and $\bar{\gamma}(0) = P_U(x,y)$
is any point in $\pi^{-1}(x)$. If f is a function on E, then differentiating it along $\bar{\gamma}$ yields, from (5.19)

$$\dot{\gamma}^a(\partial f/\partial x^a + \Gamma_a^{\ A}\ \partial f/\partial y^A) = \dot{\gamma}^a \partial f/\partial x^a + \omega_{(\gamma)}^\alpha\ X_\alpha^{\ A}(y)\ \partial f/\partial y^A$$

where $\omega_{(\gamma)}^\alpha$ are functions depending on the choice of γ, and
$X_\alpha^{\ A}\ \partial/\partial y^A$ are a basis for $\underline{G}_F$. Comparing coefficients, one finds
that

$$\omega_{(\gamma)}^\alpha = \dot{\gamma}^a\ \omega_a^\alpha$$

where ω_a^α are functions (induced on E from functions) on M; consequently the connection coefficients must admit a factorization

$$\Gamma_a^{\ A}(x,y) = \omega_a^{\ \alpha}(x)\ X_\alpha^{\ A}(y)\ . \tag{5.20}$$

The index α ranges and sums over 1, 2, ..., dim G.

To sum up: If now $E = (E, M\ \pi, F, G, H)$ is a fibre bundle
with typical fibre F, structural group G, and G-connection H,
then in adapted coordinates x^a (from the base) and y^A (from the
fibre), a basis for horizontal vector fields has the form

$$H_a = \partial/\partial x^a + \omega_a^{\ \alpha}(x)\ X_\alpha^{\ A}(y)\ \partial/\partial y^A \tag{5.21}$$

where

$$X_\alpha = X_\alpha^{\ A}\ \partial/\partial y^A \tag{5.22}$$

is a basis for the Lie algebra $\underline{G}_F$ and the $\omega_a^{\ \alpha}$ are (lifted from)

functions on the base M. The module H^* of vertical forms is
generated by

$$\theta^A = dy^A - X_\alpha{}^A (y) \; \omega^\alpha(x) \tag{5.23}$$

where

$$\omega^\alpha = \omega_a{}^\alpha(x) \; dx^a \tag{5.24}$$

are 1-forms which may also be considered (as is evident from our
abusive notation) to be lifted from M. Introducing the factorization (5.20) into (5.10), one obtains the components of the
curvature tensor in the form

$$\Omega_{ab}{}^B = (\partial\omega_b{}^\gamma/\partial x^a - \partial\omega_a{}^\gamma/\partial x^b + \omega_a{}^\alpha \; \omega_b{}^\beta \; C_{\alpha\beta}{}^\gamma) \; X_\gamma{}^B \tag{5.25}$$

where $C_{\alpha\beta}{}^\gamma$ are the structure constants of $\underline{G}$. Moreover

$$d\theta^A = (d\omega^\gamma + \tfrac{1}{2} \; C_{\alpha\beta}{}^\gamma \; \omega^\alpha \wedge \omega^B) \; X_\gamma{}^A \quad (\bmod \; \theta^B)$$

$$= \tfrac{1}{2} \; \Omega_{ab}{}^A \; dx^a \wedge dx^b . \tag{5.26}$$

If the action of G on F is effective, then there is a
standard construction of a principal fibre bundle $P = (P,M,\pi',G)$
with structure group and fibre G associated to E, and the
connection H on E induces a connection H' on P [22]. We shall
not need this construction in detail. In the applications con-
templated here, G may always be chosen so that its action is
effective, and indeed, even so that E has the additional structure
of a vector bundle with linear connection H. However, the linear

76

structure is not always the most appropriate for the applications. We shall give some examples later on. But first we explain how a connection may be associated with a Bäcklund map.

Recall that a Bäcklund map

$$\psi : J^h(M,N_1) \times N_2 \to J^1(M,N_2)$$

induces on $J^h(M,N_1) \times N_2$ a module of 1-forms $\psi^* \Omega^1(M,N_2)$, which is the pull-back of the contact module on $J^1(M,N_2)$. The forms of this module may be chosen as vertical forms defining a connection on the product space

$$(J^h(M,N_1) \times N_2, \ J^h(M,N_1), \ pr_1)$$

with fibre N_2. In a recent note we have called this connection the Bäcklund connection defined by the Bäcklund map. At this point it would have been agreeable to be able to say that the integrability conditions for the Bäcklund map were the vanishing of the curvature of the Bäcklund connection. Unhappily this is not the case; the vanishing of this curvature imposes additional conditions which one could not expect to satisfy. Alternatively, one might attempt to construct a connection, with module of vertical forms chosen to be $\widetilde{\Omega}^{h,\psi}$, on $(J^h(M,N_1) \times N_2, \ M, \ \alpha)$, or something of that kind, but $\widetilde{\Omega}^{h,\psi}$ is not quite large enough for this purpose, because of the intervention of the highest derivatives of the independent variables z^μ, which, as has already been observed, enter the formalism in a different way from the rest.

To circumvent the first difficulty, one may eliminate the additional conditions by constructing sections of $(J^h(M,N_1) \times N_2, M \times N_2, \alpha \times id)$, namely jets of maps from M to N_1, thus converting the z^μ and their derivatives into functions on the base M. To circumvent the second difficulty, one may work with the projective limit $J(M,N_1) \times N_2$, where $J(M,N_1)$ is the jet bundle of infinite order, so that there is no longer any highest derivative. We shall elaborate these two procedures in turn, but, as was remarked at the beginning of this section, we regard neither as entirely satisfactory, and suspect that there must be a more direct approach.

First of all suppose that

$$\psi : J^h(M,N_1) \times N_2 \to J^1(M,N_2). \tag{5.27}$$

is an ordinary Bäcklund map with integrability conditions

$$\widetilde{D}^{(h+1)}_c \psi^A_b - \widetilde{D}^{(h+1)}_b \psi^A_c = 0 ; \tag{5.28}$$

these are just equations (4.20). Recalling that $M \times N_2$ may always be identified with $J^o(M,N_2)$, observe that to any map

$$f : M \to N_1$$

one may associate a map

$$\widetilde{j^h f} := j^h f \times id : J^o(M,N_2) = M \times N_2 \to J^h(M,N_1) \times N_2. \tag{5.29}$$

Any (contact) form in $pr_1^* \Omega^h(M,N_1)$ is annihilated by this map.

Therefore the module $H^*_{f,\psi} := \widetilde{j^h f}^*(\widetilde{\Omega^h}, \psi)$ on $J^0(M,N_2)$ is precisely $\widetilde{j^h f}^* \psi^* \Omega^1(M,N_2)$; it has a basis

$$\theta^A = dy^A - \Gamma^A_a \, dx^a$$

where

$$\Gamma_a^{\ A} = \psi_a^{\ A} \circ \widetilde{j^h f} \tag{5.30}$$

(compare equation (4.12)). This module may be chosen as a module of vertical forms defining an horizontal distribution $H_{f,\psi}$ on the trivial fibre bundle $E = (J^0(M,N_2), M, \alpha)$. This horizontal distribution depends on the choice of both the Bäcklund map ψ and the map f. If the $\Gamma_a^{\ A}$ factorize as in (5.20) then the $X_\alpha^{\ A}$ may be identified with members of a subset of a basis for a Lie algebra $\underline{G}_{N_2}$ of vector fields acting on N_2, which may be supposed to be the Lie algebra of a Lie group G which is the structure group of the fibre bundle $E = (J^0(M,N_2), M, \alpha, N_2, G, H_{f,\psi})$. It is not justified to suppose that the $X_\alpha^{\ A}$ which actually occur in $\Gamma_a^{\ A}$ are themselves the components of a basis for $\underline{G}_{N_2}$, since some of the $\omega_a^{\ \alpha}$ might vanish.

Since the $\Gamma_a^{\ A}$ are constructed from the $\psi_a^{\ A}$ of the Bäcklund map according to (5.30) by substituting for the jet bundle variables z^μ, $z^\mu_{\ a}$, ... without altering the y^A, it follows that if the $\Gamma_a^{\ A}$ factorize for a general f, then the $\psi_a^{\ A}$ factorize, in the form

$$\psi_a^{\ A} = \tilde{\omega}_a^{\ \alpha}(x^b, z^\mu, z^\mu_{\ a}, \ldots) \, X_\alpha^{\ A}(y) \,, \tag{5.31}$$

where $\tilde{\omega}_a{}^\alpha$ are some functions pulled back by pr_1 from $J^h(M,N_1)$. Conversely, if the $\psi_a{}^A$ factorize, as in (5.31), so, trivially, do the $\Gamma_a{}^A$.

It should be pointed out that the existence of the structure group is sufficient for the ordinaryness of a Bäcklund map, although it may not be necessary. If the factorization (5.31) obtains, then the integrability conditions (5.28) for the Bäcklund map take the form

$$\tilde{\Omega}_{cb}{}^\alpha X_\alpha{}^A = 0 , \qquad (5.32)$$

where

$$\tilde{\Omega}_{cb}{}^\alpha = \tilde{D}_c^{(h+1)} \tilde{\omega}_b{}^\alpha - \tilde{D}_b^{(h+1)} \tilde{\omega}_c{}^\alpha + C_{\gamma\beta}{}^\alpha \tilde{\omega}_c{}^\gamma \tilde{\omega}_b{}^\beta \qquad (5.33)$$

and the $X_\alpha{}^A$ are among the components of basis vectors of $\underline{G}_{N_2}$ while the $C_{\gamma\beta}{}^\alpha$ are structure constants of $\underline{G}_{N_2}$.

Consequently the system of differential equations Z, introduced in the definition of a Bäcklund map at the end of section 4, is given by the vanishing of the $\tilde{\Omega}_{cb}{}^\alpha$, which are functions on $J^h(M,N_1)$, and the Bäcklund map determined by (5.31) is indeed ordinary. Moreover, the curvature of the connection $H_{f,\psi}$ has components $\widehat{j^h f}^* \tilde{\Omega}_{cb}{}^\alpha$, and so the curvature vanishes if and only if f is a solution of the integrability conditions for ψ. The vanishing of the curvature is necessary and sufficient for the integrability of the connection, in which case there exist maps

$$g : M \to N_2$$

whose graphs are local sections

$$\sigma_g : M \to J^O(M,N_2)$$ (5.34)

satisfying

$$\sigma_g{}^* \theta^A = 0 \ .$$

If in local coordinates, g is given by

$$y^A = g^A(x^b),$$ (5.35)

then

$$\frac{\partial g^A}{\partial x^a} = \Gamma_a{}^A(x)$$

where $\Gamma_a{}^A$ is given by (5.30).

Moreover, if ψ determines an ordinary Bäcklund transformation from Z to a system of differential equations $Z' \subset J^{s+1}(M,N_2)$, then Im $j^{s+1}g \subset Z'$ and g is a solution of the Bäcklund-transformed equations Z'. It is worthwhile to put these ideas into a more general context. The question, when can the Bäcklund problem be solved by an ordinary Bäcklund map, seems to require more or less the enumeration of connections with given curvature, and it appears that little can be said except in special cases. The question, does a given ordinary Bäcklund map yield a structure group, has an affirmative answer under fairly wide conditions, although it is a family of Lie groups, rather than a single one, which usually emerges. This association of a Lie group with a system of differential equations is different from the usual association of a symmetry group with such a system. We shall return to this question elsewhere.

As an application of the ideas developed in this section we describe the construction of linear scattering equations with the help of a Bäcklund map. In particular, we carry this out for the case of the sine-Gordon equation, using the Bäcklund map given in Example 3.2. In a note already published [18], M. Crampin and two of us pointed out that the linear scattering problem for a non-linear evolution equation could be described in terms of a linear connection on a principal $SL(2,\mathbb{R})$-bundle, which we called the "soliton connection". The principal bundle appearing in this description is the one associated, in the manner described just after equation (5.26) above, with the bundle whose bundle space is $J^O(M,N_2)$.

Suppose once more that a Bäcklund map

$$\psi : J^h(M,N_1) \times N_2 \to J^1(M,N_2)$$

is a solution of the Bäcklund problem for a system Z of differential equations and that the functions ψ_a^A defining ψ factorize as in (5.31), so that ψ defines a G-connection, for some Lie group G, not necessarily unique. Then by Ado's theorem ([47] p.202), the Lie algebra $\underline{G}$ of G admits a faithful linear representation, so that if the representation defined by (5.31) is not already linear, it may be replaced by a linear one. Thus there always exists a manifold N'_2 and another Bäcklund map

$$\psi' : J^h(M,N_1) \times N'_2 \to J^1(M,N'_2) \qquad (5.36)$$

defined by functions

$$\psi'^A_a = \tilde{\omega}_a^{\ \alpha} X'^A_\alpha$$

with X'^A_α satisfying the same commutation relations as the X^A_α in (5.31), but being of the form

$$X'^A_\alpha = \Gamma^A_{B\alpha} y'^B \qquad (5.37)$$

where y'^B are local coordinates on N'_2 and $\Gamma^A_{B\alpha}$ are constants.

Example 5.1: Linear scattering equations associated to the sine-Gordon equation. We use the notation of this section and of Example 3.2 to obtain the linear scattering equations of Ablowitz et al [1, 2] for the sine-Gordon equation. The Bäcklund map (3.39) induces on $J^1(M,N_1)$ a module generated by

$$\theta^1 = dy - \tilde{\omega}_a^{\ \alpha} X_\alpha^1 dx^a \qquad (\alpha = 1,2,3)$$

where

$$\tilde{\omega}_a^{\ 1} dx^a = \tfrac{1}{2} a \sin z \cdot dx^1 - \tfrac{1}{2} z_2 \, dx^2$$

$$\tilde{\omega}_a^{\ 2} dx^a = a \cos z \cdot dx^1 + a^{-1} dx^2 \qquad (5.38)$$

$$\tilde{\omega}_a^{\ 3} dx^a = - \tfrac{1}{2} a \sin z \cdot dx^1 - \tfrac{1}{2} z_2 \, dx^2$$

and $X_1^1 = 1$, $X_2^1 = y$, $X_3^1 = y^2$. The vector fields

$X_1 = d/dy$, $X_2 = yd/dy$, $X_3 = y^2 d/dy$ form a basis for a faithful one-dimensional representation of the Lie algebra $\underline{SL}(2,\mathbb{R})$, with Lie brackets

$$[X_1,X_2] = X_1, \quad [X_1,X_3] = 2X_2, \quad [X_2,X_3] = X_3 . \tag{5.39}$$

A basis for a faithful two-dimensional linear representation (5.37) with the same bracket (5.39) is given by

$$\Gamma^A_{B\ 1} = \begin{bmatrix} 0 & 0 \\ -1 & 0 \end{bmatrix}, \quad \Gamma^A_{B\ 2} = \begin{bmatrix} \frac{1}{2} & 0 \\ 0 & -\frac{1}{2} \end{bmatrix}, \quad \Gamma^A_{B\ 3} = \begin{bmatrix} 0 & -1 \\ 0 & 0 \end{bmatrix} \tag{5.40}$$

Therefore a map (5.36) with dim $N_2' = 2$ with

$$\psi'^A_{\ a} = \Gamma^A_{B\ \alpha}\, y^B\, \tilde{\omega}^\alpha_{\ a}$$

is another map solving the Bäcklund problem for the sine-Gordon equation, with induced module generated by

$$\theta^A = dy^A - \psi'^A_{\ a}\, dx^a \qquad (A = 1,2) .$$

If $f : M \to N_1$ is any solution of the sine-Gordon equation then the forms $(\widetilde{j^1 f})^* \theta^A$ comprise a basis for vertical forms on $M \times N_2'$. The curvature of the corresponding connection vanishes and therefore there exist sections $\sigma_g : M \to J^0(M,N_2')$, as described in (5.34), such that

$$\frac{\partial}{\partial x^1} \begin{bmatrix} g^1 \\ g^2 \end{bmatrix} = \frac{1}{2}\, a \begin{bmatrix} \cos f & \sin f \\ \sin f & -\cos f \end{bmatrix} \begin{bmatrix} g^1 \\ g^2 \end{bmatrix},$$

$$\frac{\partial}{\partial x^2} \begin{bmatrix} g^1 \\ g^2 \end{bmatrix} = \frac{1}{2} \begin{bmatrix} a^{-1} & -f_2 \\ f_2 & -a^{-1} \end{bmatrix} \begin{bmatrix} g^1 \\ g^2 \end{bmatrix}, \tag{5.41}$$

84

where $f_2 := \partial f/\partial x^2$. From the point of view developed in this

section these linear scattering equations are simply equations of

parallel transport (5.13) with $\Gamma_a{}^A$ given by

$$\Gamma_a{}^A = \Gamma_B{}^A{}_\alpha \, y^{B\sim}_{\omega}{}_a^\alpha.$$

In the note referred to above [18] similar formulations are given

for the KdV and modified KdV equations.

The scattering equations of Ablowitz et al have been dis-

cussed from other points of view by Corones [12], Dodd and

Gibbon [21], Hermann [43] and Morris [59, 61], for example. $\square$

Returning now to the general theory, we shall explain how a

Bäcklund map may be interpreted as defining a connection on an

infinite-dimensional manifold constructed from a jet bundle of

infinite order.

The jet bundle of infinite order $J(M,N)$ is the projective

limit (also called the inverse limit) of the bundles $J^k(M,N)$,

defined as follows ([53] p.55): Consider the infinite product

$\prod_{k=o}^{\infty} J^k(M,N)$, whose elements are sequences $\xi = (\xi_0, \xi_1, \ldots, \xi_k, \ldots)$
with $\xi_k \in J^k(M,N)$. Then $J(M,N)$ is the subset consisting of

sequences related by the natural projection:

$$J(M,N) = \{(\xi \in \prod_{k=o}^{\infty} J^k(M,N) \mid \pi^k_\ell \, \xi_k = \xi_\ell \text{ for all } k \text{ and all } \ell \leqslant k\}.$$

Consequently f and $g \in C^\infty(M,N)$ define the same point of $J(M,N)$ if

they have the same Taylor expansions, to every order, about a

point of M. There is a natural projection

$$\pi_k \; : \; J(M,N) \to J^k(M,N) \quad ,$$

for every k, by

$$(\xi_0, \; \xi_1, \; \ldots, \; \xi_k, \; \ldots) \to \xi_k \quad ,$$

with the property that

$$\pi_\ell^k \circ \pi_k = \pi_\ell \quad \text{for} \quad k \geq \ell.$$

The image of $j_x^0 f \times j_x^1 f \times \ldots \times j_x^k f \times \ldots$ in $\prod\limits_{k=0}^{\infty} J^k(M,N)$ lies in $J(M,N)$ and defines the infinite jet jf of any $f \in C^\infty(M,N)$. The ideas of differentiable function, vector field, form, and so on, generalize straightforwardly. A function f on $J(M,N)$ is differentiable if for some k there is a differentiable function f_k on $J^k(M,N)$ such that $f = f_k \circ \pi_k$. More generally, a map $\phi : J(M,N) \to P$ into a manifold P is differentiable if for some k there is a differentiable map $\phi_k : J^k(M,N) \to P$ such that $\phi = \phi_k \circ \pi_k$.

A curve γ at $\xi \in J(M,N)$ is a map

$$\gamma : I \to J(M,N)$$

where I is an open interval of $\mathbb{R}$, $\gamma(0) = \xi$, and $f \circ \gamma$ is a differentiable (real) function for every differentiable function f on $J(M,N)$. Two curves γ_1 and γ_2 at ξ are equivalent if

$$\frac{d}{dt} (f \circ \gamma_1)_{t=0} = \frac{d}{dt} (f \circ \gamma_2)_{t=0}$$

for every differentiable f. A tangent vector X at ξ is an

equivalence class of curves at ξ. If f is any (differentiable)
function on $J(M,N)$ then Xf means $\frac{d}{dt} (f \circ \gamma)_{t=0}$, where γ is any
curve in the equivalence class defining X. But there is a
function f_k on $J^k(M,N)$, for some k, such that $f = f_k \circ \pi_k$, so
that $Xf = \frac{d}{dt} (f_k \circ \pi_k \circ \gamma)_{t=0}$, which is the derivative of f_k along
the curve $\pi_k \circ \gamma$ on $J^k(M,N)$. Moreover, as is easily seen, this
derivative is independent of the choice of γ in the equivalence
class, and so by the usual arguments may be written

$$Xf = \xi^a \frac{\partial f_k}{\partial x^a} + \zeta^\mu \frac{\partial f_k}{\partial z^\mu} + \ldots + \zeta^\mu_{a_1 \ldots a_k} \frac{\partial f_k}{\partial z^\mu_{a_1 \ldots a_k}} \qquad (5.42)$$

where all the partial derivatives of f_k are evaluated at $\pi_k(\xi)$.

A vector field is defined, pointwise, to be an assignment of
vectors which takes differentiable functions into differentiable
functions.

If X is a vector field on $J(M,N)$ and f_k is any function on
$J^k(M,N)$ then for some $\ell(k) \geqslant k$ there is a function g_ℓ on $J^\ell(M,N)$
such that $X(f_k \circ \pi_k) = g_\ell \circ \pi_\ell$, so that the action of X on f_k is
given by

$$X(f_k \circ \pi_k) = [(\xi^a \frac{\partial}{\partial x^a} + \zeta^\mu \frac{\partial}{\partial z^\mu} + \zeta^\mu_a \frac{\partial}{\partial z^\mu_a} + \ldots$$

$$+ \zeta^\mu_{a_1 \ldots a_k} \frac{\partial}{\partial z^\mu_{a_1 \ldots a_k}}) f_k \circ \pi^\ell_k] \circ \pi_\ell$$

where $\xi^k, \zeta^\mu, \ldots, \zeta^\mu_{a_1 \ldots a_k}$ are functions on $J^\ell(M,N)$ independent
of the choice of f_k. Thus the action of X may be specified by
writing

$$X = \xi^a \frac{\partial}{\partial x^a} + \zeta^\mu \frac{\partial}{\partial z^\mu} + \zeta^\mu_a \frac{\partial}{\partial z^\mu_a} + \ldots + \zeta^\mu_{a_1 \ldots a_k} \frac{\partial}{\partial z^\mu_{a_1 \ldots a_k}} + \ldots$$

where all of the coefficients ξ^a, ζ^μ, $\ldots$ are differentiable

functions on $J(M,N)$.

Let χ denote the collection of vector fields on $J(M,N)$.

Then a p-form ω on $J(M,N)$ is a skew-symmetric $C^\infty(J(M,N))$-

multilinear map

$$\underbrace{\chi \times \chi \times \ldots \times \chi}_{p \text{ factors}} \to C^\infty(J(M,N)).$$

Thus $X \lrcorner \omega$ may be defined in the usual way.

If X is a vector at $\xi \in J(M,N)$ then the vector $\pi_{k*}X(\xi)$ at

$\pi_k(\xi) \in J^k(M,N)$ is defined by

$$(\pi_{k*}X)f_k = X(f_k \circ \pi_k)$$

for all $f_k \in C^\infty(J^k(M,N))$. If ω_k is a p-form on $J^k(M,N)$ then

$\pi_k^* \omega_k$ is the p-form on $J(M,N)$ defined pointwise by

$(\pi_k^* \omega_k)(X_1, \ldots, X_p) = \omega_k(\pi_{k*}X_1, \ldots, \pi_{k*}X_p)$. Thus the contact

module $\Omega(M,N)$ is the module of 1-forms on $J(M,N)$ generated by

$\sum\limits_{k=0}^{\infty} \pi_k^* \; \Omega^k(M,N)$.

All these definitions generalize in a rather obvious way to

$J(M,N_1) \times N_2$. For example, if

$$\tilde{\pi}_k = \pi_k \times \text{id} \; ,$$

where id is the identity map of N_2, then a differentiable function

f on $J(M,N_1) \times N_2$ is one of the form $f_k \circ \tilde{\pi}_k$, where f_k is a differentiable function on $J^k(M,N_1) \times N_2$.

A vector X at $\xi \in J(M,N_1) \times N_2$ is a derivation of differentiable functions. The components of a vector may be specified as in (5.42) above, with the addition of a term of the form $\eta^A \frac{\partial}{\partial y^A}$ (where y^A are local coordinates on N_2), and projections of vectors and pullbacks of forms are defined by trifling modifications of the above definitions. Then if

$$\psi : J^h(M,N_1) \times N_2 \to J^1(M,N_2)$$

is a Bäcklund map, a module of 1-forms Ω^ψ on $J(M,N)$ is generated by $\sum_{k=h}^{\infty} \tilde{\pi}_k{}^* \tilde{\Omega}^{k,\psi}$. A straightforward computation shows that if

$$X \lrcorner \Omega^\psi = 0,$$

then X is a vector field of the form $X = \xi^a \tilde{D}_a$, where ξ^a are functions on $J(M,N)$ and

$$\tilde{D}_a = \tilde{\pi}_h{}^* \psi_a^A \frac{\partial}{\partial y^A} + \frac{\partial}{\partial x^a} + z_a^\mu \frac{\partial}{\partial z^\mu} + \ldots + z_{ab_1 \ldots b_k}^\mu \frac{\partial}{\partial z^\mu_{b_1 \ldots b_k}} + \ldots$$

Thus by extending the definition of a connection to infinite jet bundles, one may define a connection H_ψ on $(J(M,N_1) \times N_2, \mathrm{pr}_1 \circ \pi_0, M)$ with $\tilde{D}_a$ as basis for horizontal vector fields and the forms in Ω^ψ as vertical forms. Since for any $f \in C^\infty(J(M,N))$,

$$[\tilde{D}_a, \tilde{D}_b] f := (\tilde{D}_a \tilde{D}_b - \tilde{D}_b \tilde{D}_a) f = (\tilde{D}_a \tilde{\pi}_h{}^* \psi_b^A - \tilde{D}_b \tilde{\pi}_h{}^* \psi_a^A) \frac{\partial}{\partial y^A} f ,$$

it is natural to identify (the vertical part of) $[\tilde{D}_a, \tilde{D}_b]$ as the curvature of this connection. <u>The vanishing of the curvature</u> <u>is then exactly the system of integrability conditions for the</u> <u>Bäcklund map</u>. This formulation has certain advantages which go with its generality and is occasionally convenient for calculations.

Section 6

ONE PARAMETER FAMILIES OF BÄCKLUND MAPS

It has been known for a hundred years that a geometrical structure
may fruitfully be studied in terms of its linearizations and its
invariance group. Here the question presents itself in the form:
how may a Bäcklund map be deformed, and what are its symmetries?
Recently, Ibragimov and Anderson [46] have studied infinitesimal
diffeomorphisms of jet bundles which preserve the contact module.
Leaving to the future a systematic study of symmetries of
Bäcklund problems we here show how extended point transformations
of jet bundles can be combined with a Bäcklund map to give a
family of Bäcklund maps with the same integrability conditions.
In particular we give conditions sufficient for a deformation of
a Bäcklund map to be an ordinary Bäcklund map with the same
integrability conditions. The deformation is constructed from
suitably related 1-parameter groups of transformations of the
independent variables x^a, the original dependent variables z^μ
and the new variables y^A. The parameter (eigenvalue) appearing
in the linear scattering equations derivable from Bäcklund maps
(as in example 5.1) may be introduced by deforming Bäcklund maps
in this way.

In the applications, a 1-parameter family of Bäcklund maps
is derived from a single one by composing it with space-time
transformations which leave the integrability conditions invariant.
For example, as is well known, the auto-Bäcklund transformation
of the sine-Gordon equation may be obtained from the Bianchi
transformation by a boost [24], and it is shown below that the
parameter appearing in the Bäcklund transformation of the Korteweg-
deVries equation may be introduced, in the same way, by a
Galilean transformation [68].

First of all we describe the prolongation of (local) diffeomorphisms from $M \times N = J^o(M,N)$ to any $J^\ell(M,N)$. The notation and
conventions are as before; in particular, α and β denote the
source and target projections. This idea of prolongation is a
slight generalization of that described in section 2, but to put
them both into the same context would require additional machinery, which we prefer to avoid.

Let ϕ be a diffeomorphism of M. A diffeomorphism ϕ^o of
$J^o(M,N)$ is said to be compatible with ϕ if it preserves fibres
of $\alpha : J^o(M,N) \to M$, that is to say, if

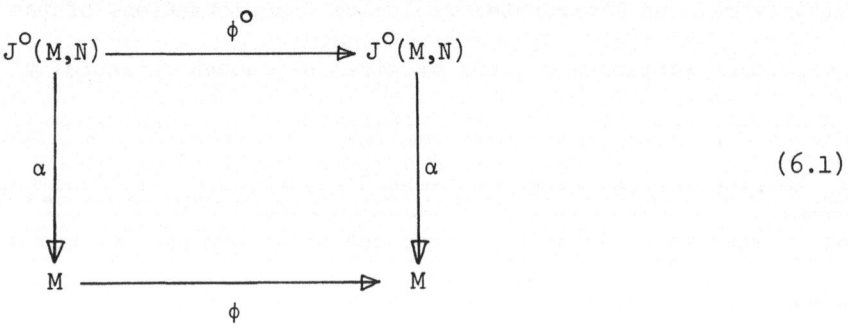

$$(6.1)$$

commutes. Whenever this is the case, the action of ϕ^o may be
extended to $C^\infty(M,N)$: if $f \in C^\infty(M,N)$, the map f^ϕ induced from f
by ϕ^o is

$$f^\phi = \beta \circ \phi^o \circ j^o f \circ \phi^{-1} . \tag{6.2}$$

Consequently the action may be extended to $J^\ell(M,N)$: let
$\xi \in J^\ell(M,N)$, and let f be any map in the equivalence class ξ, so
that $\xi = j^\ell_{\alpha(\xi)} f$. Then define the ℓth prolongation of ϕ^o,

$$\phi^\ell : J^\ell(M,N) \to J^\ell(M,N)$$

by

$$\phi^\ell(\xi) = j^\ell_{\phi\circ\alpha(\xi)} f^\phi . \tag{6.3}$$

Since ξ determines the derivatives of f at $\alpha(\xi)$, up to the ℓth, and
$j^\ell_{\phi\circ\alpha(\xi)} f^\phi$ depends on these and no more, $\phi^\ell(\xi)$ is independent of
the choice of f in the equivalence class ξ. Moreover, for any
$f \in C^\infty(M,N)$, $\phi^\ell \circ j^\ell f (x) = j^\ell f^\phi(\phi(x))$, so that ϕ^ℓ preserves the
contact module:

$$\phi^{\ell*}\Omega^\ell(M,N) = \Omega^\ell(M,N). \tag{6.4}$$

We now extend this idea to the context of Bäcklund maps:
again let ϕ be a diffeomorphism of M, and let ϕ_1^o and ϕ_2^o be
diffeomorphisms of $J^o(M,N_1)$ and $J^o(M,N_2)$ respectively, each com-
patible with ϕ. Then a diffeomorphism $\tilde{\phi}^\ell$ of $J^\ell(M,N_1) \times N_2$, also
called a prolongation of $(\phi, \phi_1^o, \phi_2^o)$, is defined uniquely by

the demand that its action on $J^\ell(M,N_1)$ be that of $\phi_1{}^\ell$ and that
its action on $M \times N_2 = J^0(M,N_2)$ be that of $\phi_2{}^0$, namely that

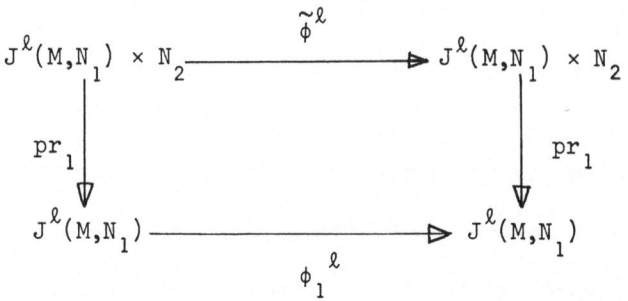

and   (6.5)

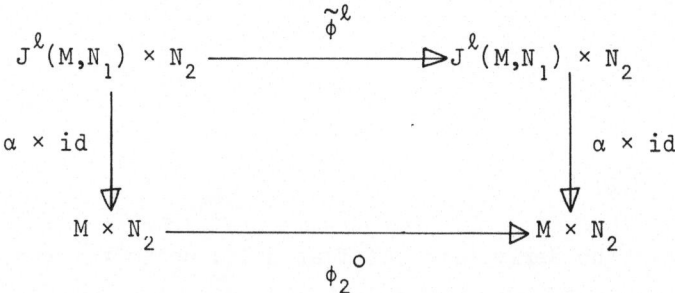

commute.

Now let $\psi : J^h(M,N_1) \times N_2 \to J^1(M,N_2)$ be an ordinary
Bäcklund map with integrability conditions $\tilde{Z}$ such that
$Z \times N_2 \subset \tilde{Z}$, where Z is a system of differential equations on
$J^{h+1}(M,N_1)$. Then the possibility arises that

$$\psi^\phi := (\phi_2{}^1)^{-1} \circ \psi \circ \tilde{\phi}^h : J^h(M,N_1) \times N_2 \to J^1(M,N_2) \qquad (6.6)$$

may also be an ordinary Bäcklund map and that it may have the
same integrability conditions as ψ. This will certainly be the
case if $\phi_1{}^{h+1} : J^{h+1}(M,N_1) \to J^{h+1}(M,N_1)$, the prolongation of ϕ
and $\phi_1{}^0$, satisfies the symmetry conditions

$$\phi_1^{h+1*}\Sigma \subset \Sigma , \qquad\qquad (6.7)$$

where Σ is the finitely generated ideal of functions in $C^\infty(J^{h+1}(M,N_1), \mathbb{R})$ with zero set Z. In this case we shall say that ϕ, ϕ_1^O, ϕ_2^O define a symmetry of ψ.

A symmetry of ψ is always defined by the maps $\phi = id_M$, $\phi_1^O = id_{J^O(M,N_1)}$, ϕ_2^O, where ϕ_2^O is any diffeomorphism of $J^O(M,N_2)$ leaving M pointwise fixed. This (trivial) symmetry corresponds to the freedom always available in choosing local coordinates y^A on N_2 and associated coordinates y^A, $y^A_{a_1} \ldots y^A_{a_1 \ldots a_\ell}$ on $J^\ell(M,N_2)$, $\ell \geqslant 0$.

Next we define a Bäcklund automorphism – the "auto-Bäcklund transformation" of several well-known examples – which will be used in section 7: if $J^O(M,N_1)$ and $J^O(M,N_2)$ are related by the identity diffeomorphism

$$(id)^O : J^O(M,N_1) \to J^O(M,N_2),$$

and the ordinary Bäcklund map ψ determines a Bäcklund transformation between systems of equations Z on $J^{h+1}(M,N_1)$ and Z' on $J^{h+1}(M,N_2)$ in such a way that

$$(id)^{h+1}(Z) = Z' ,$$

then ψ is called a Bäcklund automorphism and the transformation is called a Bäcklund self-transformation.

We next consider in further detail deformations of Bäcklund

maps which arise when the maps ϕ, $\phi_1^{\,o}$ and $\phi_2^{\,o}$ all belong to
1-parameter groups of diffeomorphisms and (6.7) is satisfied by
maps ϕ_1^{h+1} other than the identity.

Let ϕ_t be a 1-parameter group of transformations of M
($t \in \mathbb{R}$). Then a 1-parameter group $\phi_t^{\,o}$ of transformations of
$J^o(M,N)$ is compatible with ϕ_t if $\phi_t^{\,o}$ is compatible with ϕ_t for
each t. When this is the case, the ℓth prolongations $\phi_t^{\,\ell}$ of $\phi_t^{\,o}$
comprise a 1-parameter group of transformations of $J^\ell(M,N)$,
called the ℓth prolongation of $\phi_t^{\,o}$.

If $X = \xi^b \partial_b$ is the vector field generating ϕ_t, then for
compatibility the vector field generating $\phi_t^{\,o}$ must have the form
$X^o = \zeta^\mu \partial/\partial z^\mu + (\xi^b \circ \alpha)\partial_b$, where ζ^μ are functions on $J^o(M,N)$.
The vector field X^ℓ generating $\phi_t^{\,\ell}$ is called the ℓth prolongation
of X^o; from (6.4) it has the property

$$\pounds_{X^\ell} \, \Omega^\ell(M,N) \subset \Omega^\ell(M,N) \; , \tag{6.8}$$

where $\pounds_{X^\ell}$ denotes the Lie derivative with respect to X^ℓ.
A straightforward calculation from (6.3) shows that

$$X^\ell = \pi_\ell^{\ell+1}{}_* \, [\xi^c \, D_c^{(\ell+1)} + \sum_{r=o}^{\ell} D_{b_1 \ldots b_r}^{(\ell+1)} \, (\zeta^\mu - z^\mu{}_c \, \xi^c)] \tag{6.9}$$

(where ξ^b and ζ^μ have, abusively, been written for $\xi^c \circ \alpha$ and
$\zeta^\mu \circ \pi_o^{\ell+1}$).

Now let ϕ_t be a 1-parameter group of transformations of M,
and let $\phi_{t1}^{\,o}$ and $\phi_{t2}^{\,o}$ be 1-parameter groups of transformations
of $J^o(M,N_1)$ and $J^o(M,N_2)$ respectively, each compatible with ϕ_t.

Then the prolongations $\tilde{\phi}_t{}^\ell$ of $(\phi_t, \phi_{t1}{}^o, \phi_{t2}{}^o)$, defined for each

t by (6.5), comprise a 1-parameter group of transformations of

$J^\ell(M,N_1) \times N_2$.

Let X denote the vector field generating $\phi_{t1}{}^{h+1}$ on

$J^{h+1}(M,N_1)$. Then X acts on the subset Z of $J^{h+1}(M,N_1)$. If

(locally) $\phi_{t1}{}^{h+1}$ and hence X, may be chosen so that

$$\mathcal{L}_X Z \subset Z , \tag{6.10}$$

then (compare (6.6)) the maps

$$\psi_t{}^\phi := \phi_{-t2}{}^1 \circ \psi \circ \tilde{\phi}_t{}^h : J^h(M,N_1) \times N_2 \to J^1(M,N_2) \tag{6.11}$$

comprise a 1-parameter family of ordinary Bäcklund maps, with

$\psi_o{}^\phi = \psi$, all having integrability conditions $\tilde{Z}$ with $Z \times N_2 \subset \tilde{Z}$.

The maps $\psi_t{}^\phi$ will be called deformations of ψ.

In practice, and in the examples below, (6.10) is regarded,

for a given Bäcklund problem, as a set of conditions on X to be

satisfied, if possible, using (6.9). This determines ϕ_t and

$\phi_{t1}{}^o$. The choice of $\phi_{t2}{}^o$ is motivated by considerations particular

to the given problem. In the interesting cases the module $\tilde{\Omega}^{\ell,\psi}$

is not invariant under the group action.

Example 6.1. The sine-Gordon equation again: Pursuing Example

3.1, set a = 1 in (3.34), to obtain the "Bianchi transformation"

ψ:

$$y_1' = z_1 + 2 \sin\tfrac{1}{2}(y + z),$$
$$\tag{6.12}$$
$$y_2' = - z_2 + 2 \sin\tfrac{1}{2}(y - z).$$

The integrability conditions

$$z_{12} - \sin z = 0 \qquad\qquad (3.37)$$

define a submanifold Z of $J^2(M, N_1)$ which is invariant under the action of the vector fields

$$X_1 = x^1 \frac{\partial}{\partial x^1} - x^2 \frac{\partial}{\partial x^2} - z_1 \frac{\partial}{\partial z_1} + z_2 \frac{\partial}{\partial z_2} - 2z_{11} \frac{\partial}{\partial z_{11}} + 2z_{22} \frac{\partial}{\partial z_{22}} \, ,$$

$$X_2 = \frac{\partial}{\partial x^1} \, , \quad \text{and}$$

$$X_3 = \frac{\partial}{\partial x^2} \, .$$

We shall calculate the deformations of ψ by the 1-parameter group which X_1 generates. The vector fields X_2 and X_3 generate only the identity deformation and are therefore not of present interest.

It is easy to check that X_1 generates a prolongation to $J^2(M, N_1)$ of the 1-parameter group of transformations $\phi_{t1}{}^\circ$ of $J^\circ(M, N_1)$ whose action is given by

$$\phi_{t1}{}^\circ : (x^1, x^2, z) \mapsto (e^t x^1, \ e^{-t} x^2, z) \, ,$$

and that $\phi_{t1}{}^\circ$ is compatible with the 1-parameter group of transformations ϕ_t of M whose action is given by

$$\phi_t : (x^1, x^2) \mapsto (e^t x^1, \ e^{-t} x^2) \, .$$

The transformations ϕ_t, originally introduced in this context by Lie [55], are called transformations of Lie, or, in the language of special relativity theory, boosts. A 1-parameter group of transformations $\phi_{t2}{}^{\circ}$ of $J^{\circ}(M,N_2)$ compatible with ϕ_t may be constructed by choosing its action on $J^{\circ}(M,N_2)$ to be identical with that of $\phi_{t1}{}^{\circ}$ on $J^{\circ}(M,N_1)$, namely

$$\phi_{t2}{}^{\circ}: (x^1,x^2,y) \mapsto (e^t x^1, e^{-t} x^2, y). \tag{6.13}$$

Such a choice may always be made when $\dim N_1 = \dim N_2$.

The prolongation $\tilde{\phi}_t{}^1$ to $J^1(M,N_1) \times N_2$ is given by

$$\tilde{\phi}_t{}^1: (x^1,x^2,z,z_1,z_2,y) \mapsto (e^t x^1, e^{-t} x^2, z, e^{-t} z_1, e^t z_2, y). \tag{6.14}$$

Substituting from (6.12), (6.13) and (6.14) into (6.11) one obtains deformations of the Bianchi transformation ψ to the Bäcklund transformation ψ_t:

$$y_1 = 2e^t \sin\tfrac{1}{2}(y + z) + z_1,$$

$$y_2 = 2e^{-t} \sin\tfrac{1}{2}(y - z) - z_2.$$

This is precisely the transformation (3.34), with e^t written for a, and shows explicitly how the Bäcklund transformation may be obtained from the Bianchi transformation by a boost [24]. □

Example 6.2 The KdV potential equation again: Pursuing example 4.1, write ψ for the map defined by setting $t = 0$ in (4.38):

$$y_1' = \psi_1(0) = - z_1 + (y - z)^2,$$

$$y_2' = \psi_2(0) = - z_2 + 4 \{z_1^2 + z_1(y - z)^2 + z_{11}(y - z)\}. \tag{6.15}$$

The l-parameter group of Galilean boosts of M is

$$\phi_t : (x^1, x^2) \mapsto (x^1 - 12t\, x^2, x^2). \qquad (6.16)$$

There is a unique l-parameter group of transformations $\phi_{t1}{}^o$ of $J^o(M, N_1)$ compatible with ϕ_t, and such that the KdV potential equation is invariant under the prolongation $\phi_{t1}{}^3$, namely

$$\phi_{t1}{}^o : (x^1, x^2, z) \mapsto (x^1 - 12tx^2, x^2, z + tx^1 - 6t^2x^2). \qquad (6.17)$$

The action of $\phi_{t1}{}^3$ yields

$$z_1 \mapsto z_1 + t$$

$$z_2 \mapsto z_2 + 12tz_1 + 6t^2 \qquad (6.18)$$

$$z_{111} \mapsto z_{111}$$

and the invariance of $z_{111} + z_2 - 6z_1{}^2$, whose vanishing yields the KdV potential equation, follows at once from (6.17) and (6.18).

A l-parameter group of transformations $\phi_{t2}{}^o$ of $J^o(M, N_2)$, compatible with ϕ_t, is given by an action identical to that of

$$\phi_{t2}{}^o : (x^1, x^2, y) \mapsto (x^1 - 12t\, x^2, x^2, y + tx^1 - 6t^2\, x^2). \qquad (6.19)$$

From (6.17) – (6.18) one may calculate the action of the prolongation $\tilde{\phi}_t{}^2$ on $J^2(M, N_1) \times N_2$, but in addition to these formulae we need only

$$z_{11} \mapsto z_{11}. \qquad (6.20)$$

Substituting from (6.17) - (6.20) in (6.11), the equation for a deformation, one finds that

$$\psi_t^{\phi} = \phi_{-t2}^{1} \circ \psi \circ \phi_t^2$$

is precisely the Estabrook–Wahlquist Bäcklund map (4.38) of the KdV potential equation from which we started. This example, which is a reformulation of a result of Steudel [68], shows how the parameter t in the map (4.38) arises from the Galilean boost (6.16). (cf. example 7.2 below). □

Example 6.3 The KdV and modified KdV equations again. In the preceding two examples, the Bäcklund transformation determined by the Bäcklund map is an "auto-Bäcklund" transformation, or Bäcklund self-transformation, which, if N_1 and N_2 are identified, is seen actually to leave the differential equation Z invariant; that is to say, in each of these cases, $Z' = Z$. In such cases, Z' and Z of course have exactly the same symmetry properties, but in general this is not so. We now exhibit a 1-parameter family of Bäcklund maps, obtained from the Galilean invariance of the KdV equation, which yield the Galilean-boosted forms of the modified KdV equation; the latter is not Galilean-invariant.

Pursuing example 4.2, denote the map (4.43) by ψ:

$$y_1' = \psi_1 := -2z - y^2$$

$$y_2' = \psi_2 := 8z^2 + 4zy^2 + 2z_{11} - 4z_1 y$$

(6.21)

Again let ϕ_t denote the 1-parameter group of Galilean boosts of M, as in (6.16). Then 1-parameter groups of transformations of $J^o(M,N_1)$ and $J^o(M,N_2)$, compatible with ϕ_t, are given respectively by

$$\phi_{t1}^{\ o} : (x^1,x^2,z) \mapsto (x^1 - 12tx^2, x^2, z - t)$$

and   (6.22)

$$\phi_{t2}^{\ o} : (x^1,x^2,y) \mapsto (x^1 - 12tx^2, x^2, y) .$$

It is easily verified that the prolongation $\phi_{t1}^{\ 3}$ to $J^3(M,N_1)$ has the effect

$$z \mapsto z - t$$

$$z_1 \mapsto z_1$$

$$z_2 \mapsto z_2 + 12tz_1$$ (6.23)

$$z_{111} \mapsto z_{111}$$

so that

$$z_{111} + z_2 + 12zz_1 \mapsto z_{111} + z_2 + 12zz_1 \ ;$$

thus the KdV equation is invariant [37]. To compute the action of the deformed map we need besides (6.22) and (6.23) only

$$z_{11} \mapsto z_{11}$$

and may then compute [72] $\psi_t^{\phi} = \phi_{-t2}^{\ 1} \circ \psi \circ \tilde{\phi}_t^{\ 2}$, which is given by

$$y_1 = -2z - y^2 + 2t,$$

$$(6.24)$$

$$y_2 = 4(z + 2t)(2z + y^2 - 2t) + 2z_{11} - 4z_1 y.$$

The image of the KdV equation under (6.24) is the zero set of

$$y_{111} + y_2 - 6y^2 y_1 + 12t y_1$$

that is to say, the Galilean-boosted form of (4.44).   □

Section 7

SOLUTIONS OF THE BÄCKLUND PROBLEM

In previous sections it has been assumed that a Bäcklund map is
given, and taken for granted that its integrability conditions
are interesting. We turn now to the main practical problem which
is to find Bäcklund maps whose integrability conditions are a
given system of partial differential equations. On the face of
it, the solution to this problem, which we have called the
Bäcklund problem, requires the determination of functions $\psi_a{}^A$
such that $\widetilde{D}_{[a}^{(h+1)} \psi^A{}_{b]} = 0$ (equations (4.20)) are combinations of
the given equations. In practice, only special solutions of this
problem have been found, and there is no obvious criterion for
deciding which solutions are interesting ones. However, Wahlquist
and Estabrook [72] have invented a very efficient way of finding
Bäcklund transformations, which in effect yields solutions of the
Backlund problem, although not in the language of jet bundles
used here. Their method employs Cartan's theory of exterior
differential systems [8].

In this section we explain the relationship of their method
to the jet bundle formalism, generalize it to the case of any
number of independent variables, and by way of example treat the

104

sine-Gordon equation and extend the results of Wahlquist and
Estabrook on the Korteweg-deVries equation. Further details may
be found in the papers just referred to and in Hermann's
treatise [44].

We show first of all how a quasi-linear system of differential equations Z on $J^{h+1}(M,N_1)$ may be replaced by an exterior
differential system I_Z on $J^h(M,N_1)$ which has the same solutions.
It has already been observed, from inspection of (4.8) and (4.9),
that the integrability conditions for Bäcklund maps of the kind
considered in this paper are always quasi-linear.

Recall that for $\ell = 0,\ldots, h$ a basis for the contact ideal
consists of the 1-forms

$$\theta^\mu_{a_1\ldots a_\ell} = dz^\mu_{a_1\ldots a_\ell} - z^\mu_{a_1\ldots a_\ell b}\, dx^b\,.$$

Let $\varepsilon_{b_1 b_2 \ldots b_m}$ denote the m-dimensional Levi-Civita symbol,
and construct

$$\varepsilon_{cb_2\ldots b_m}\, \theta^\mu_{a_1\ldots a_\ell} \wedge dx^{b_2} \wedge \ldots \wedge dx^{b_m}$$

$$= \varepsilon_{cb_2\ldots b_m}\, dz^\mu_{a_1\ldots a_\ell} \wedge dx^{b_2} \wedge \ldots \wedge dx^{b_m}$$

$$\quad - z^\mu_{a_1\ldots a_\ell b}\, \varepsilon_{cb_2\ldots b_m}\, dx^b \wedge dx^{b_2} \wedge \ldots \wedge dx^{b_m}$$

$$= \varepsilon_{cb\ldots b_m}\, dz^\mu_{a_1\ldots a_\ell} \wedge dx^{b_2} \wedge \ldots \wedge dx^{b_m}$$

$$\quad - (m-1)!\, z^\mu_{a_1\ldots a_\ell c}\, \omega \tag{7.1}$$

where

$$\omega = (m!)^{-1} \, \varepsilon_{b_1 \ldots b_m} \, dx^{b_1} \wedge \ldots \wedge dx^{b_m} . \tag{7.2}$$

Any quasi-linear function F on $J^{h+1}(M,N_1)$ has the form

$$F = A_\mu{}^{a_1 \ldots a_{h+1}} \, z^\mu{}_{a_1 \ldots a_{h+1}} + B \tag{7.3}$$

where B and all the $A_\mu{}^{a_1 \ldots a_{h+1}} = A_\mu{}^{(a_1 \ldots a_{h+1})}$ are (lifts of) functions on $J^h(M,N_1)$. Now any jet extension $j^{h+1} f$ annihilating F must annihilate Fω, and conversely, since $j^{h+1} f$ is a section of the source map. Moreover $j^{h+1} f$ annihilates the contact ideal $\Omega^{h+1}(M,N_1)$, so that $(j^{h+1} f)^*$ (right hand side of (7.1)) = 0.

Therefore $(j^{h+1} f)^* F = 0$ iff $(j^h f)^* F' = 0$, where

$$F' = A_\mu{}^{a_1 \ldots a_h c} \, \varepsilon_{cb_2 \ldots b_m} \, dz^\mu{}_{a_1 \ldots a_h} \wedge dx^{b_2} \ldots \wedge dx^{b_m} + (m-1)! \, B\omega \tag{7.4}$$

is an m-form on $J^h(M,N_1)$. Note that in the case where there is a fixed set of indices $\{c_1, \ldots, c_{h+1}\}$ (we suspend for the moment the range convention for these indices) such that F has the form

$$F = z^\mu{}_{c_1 \ldots c_{h+1}} + B$$

then $(j^{h+1}f)^* F = 0$ iff $(j^h f)^* F'_{c_k} = 0$ $\quad k = 1, \ldots, h + 1$

where

$$F'_{c_k} := \varepsilon_{c_k b_2 \ldots b_m} \, dz^\mu{}_{c_1 \ldots \widehat{c}_k \ldots c_{h+1}} \wedge dx^{b_2} \ldots \wedge dx^{b_m} + (m-1)! \, B\omega$$

(see example 7.1 below). If θ is any k-form, let

106

$$\varepsilon_{a_1 \ldots a_k}(\theta) := \varepsilon_{b_{k+1} \ldots b_m a_1 \ldots a_k} \theta \wedge dx^{b_{k+1}} \wedge \ldots \wedge dx^{b_m}$$

and if Ω is any collection of k-forms, let

$$\varepsilon(\Omega) := \{\varepsilon_{a_1 \ldots a_k}(\theta) \mid \theta \in \Omega\} \quad .$$

Notice in particular that $(j^h f)^* \varepsilon(\theta^\mu_{a_1 \ldots a_\ell}) = 0$ (for $\ell < h$).

Now let $F_1 \ldots F_q$ be functions on $J^{h+1}(M,N_1)$ generating the ideal Σ_Z which defines a quasi-linear system of differential equations Z, and let $F'_1, \ldots, F'_q$ be the corresponding m-forms on $J^h(M,N_1)$, constructed from (7.3) by (7.4). Let I_Z be the differential ideal on $J^h(M,N_1)$ generated by $F'_1, \ldots, F'_q$ and $\varepsilon(\Omega^h(M,N_1))$. It follows from the above remarks that $f : M \to N_1$ is a solution of the system Z if and only if $j^h f$ is a solution of the exterior system I_Z, that is to say, if and only if $(j^h f)^* I_Z = 0$, and that if $\sigma_Z : M \to J^h(M,N_1)$ is any section of the source map $\alpha : J^h(M,N_1) \to M$ which annihilates the ideal I_Z, then there is a map $f : M \to N_1$ which is a solution of Z, such that $j^h f = \sigma_Z$.

The key to the transition to the Wahlquist-Estabrook procedure is that one may be able to characterise solutions of Z in terms of proper sub-ideals of I_Z.

Let I'_Z be a closed subideal of I_Z, and let $C(I_Z')$ denote the characteristic distribution of I_Z' : $X \in C(I_Z')$ iff $X \lrcorner I_Z' = 0$. We assume that $C(I_Z')$ is differentiable; in the applications it is even real-analytic. Since I_Z' is closed,

$C(I_Z')$ is completely integrable and (locally) $J^h(M,N_1)$ is foliated by leaves of $C(I_Z')$. Taking the quotient by this foliation one may construct a manifold B of dimension $(\dim J^h(M,N_1) - \dim C(I_Z'))$ and a map $\pi : J^h(M,N_1) \to B$ such that for $p \in J^h(M,N_1)$, $\pi^{-1}(\pi(p))$ is a leaf of C_Z. According to Cartan (cf. [38] theorem 4.7), it then follows that there is a closed ideal of forms I' on B such that $\pi^* I' = I_Z'$. We assume also that $\alpha_* C(I_Z') = 0$, where α is the source projection. It follows that there is a surjective map $p : B \to M$. It may occur, moreover, that every section $\rho : M \to B$ of the bundle (B,M,p) which satisfies $\rho^* I' = 0$ is of the form $\rho = \pi \circ j^h f$, where $f : M \to N$ is a solution of the system Z, in which case I' will be called effective (for Z).

The problem of finding solutions of Z is thus transformed into the problem of finding maps which annihilate effective ideals I', and the Bäcklund problem may correspondingly be transformed: let ζ be a map

$$\zeta : B \times N_2 \to J^1(M,N_2)$$

such that $\pi_0^1 \circ \zeta = p \times \mathrm{id}_{N_2}$ where N_2 is some manifold.

Suppose further that there is an ideal I' on B, effective for Z, such that

$$\zeta^* \varepsilon(d\Omega^1(M,N_2)) \subset \mathrm{pr}_1^* I' + I(\zeta^* \Omega^1(M,N_2)). \tag{7.5}$$

Now define $\psi : J^h(M,N_1) \times N_2 \to J^1(M,N_2)$ by

$$\psi = \zeta \circ (\pi \times id_{N_2}) .$$

It follows that

$$\psi^* \ \epsilon(d\Omega^1(M,N_2)) \subset pr_1{}^* I_Z{}' + I(\psi^* \Omega^1(M,N_2)) ,$$

and from the definition of an effective ideal above, comparison with (4.23) shows that (7.5) is the condition that ψ should be a solution of the Bäcklund problem for Z.

The procedure of Wahlquist and Estabrook, which, extending Cartan's use of the word, they call "prolongation", is

(1) to choose an effective ideal I',

(2) to find a manifold N_2 and a map ζ satisfying (7.5), whereupon, in view of the foregoing discussion, they have in effect solved the Bäcklund problem for Z. The remarkable efficiency of their procedure may be attributed to the reduction of dimension achieved by working with B instead of with $J^h(M,N_1)$. However, the advantages of their procedure are limited by the fact that it gives one little opportunity to observe that there may be several distinct ideals I' yielding several distinct solutions of the Bäcklund problem.

We illustrate all this by treating the sine-Gordon equation.

Example 7.1. The sine-Gordon equation. Here m = 2. The equation is defined (Example 2.2) by the single function (2.15):

$$F = z_{12} - \sin z .$$

The corresponding 2-forms are

$$F'_1 = dz_2 \wedge dx^2 - \sin z \, dx^1 \wedge dx^2$$

and $F'_2 = - dz_1 \wedge dx^1 - \sin z \, dx^1 \wedge dx^2.$

The ideal I_Z is generated by F'_1 and F'_2 together with

$$\theta \wedge dx^1 = dz \wedge dx^1 + z_2 dx^1 \wedge dx^2$$

and $\theta \wedge dx^2 = dz \wedge dx^2 - z_1 dx^1 \wedge dx^2.$

Three effective closed subideals I'_{21}, I'_{22}, I'_{23} are described in terms of generators as follows:

$$I'_{21} = \{dz \wedge dx^2 - z_1 dx^1 \wedge dx^2, \; dz \wedge dx^1 + z_2 dx^1 \wedge dx^2,$$

$$\frac{1}{2}(dz_1 \wedge dx^1 - dz_2 \wedge \dot{dx}^2) + \sin z \, dx^1 \wedge dx^2\},$$

$$I'_{22} = \{dz \wedge dx^2 - z_1 dx^1 \wedge dx^2, \; dz_1 \wedge dx^1 + \sin z \, dx^1 \wedge dx^2\},$$

$$I'_{23} = \{dz \wedge dx^1 + z_2 dx^1 \wedge dx^2, \; dz_2 \wedge dx^2 - \sin z \, dx^1 \wedge dx^2\}.$$

The corresponding characteristic distributions are

$$C(I'_{21}) = \{0\} \quad,$$

$$C(I'_{22}) = \{\partial/\partial z_2\} \quad,$$

$$C(I'_{23}) = \{\partial/\partial z_1\} \quad.$$

The corresponding manifolds B_i and maps π_i ($i = 1, 2, 3$) are as follows: The manifold B_1 may be identified with $J^1(M,N_1)$ and

110

I_1' with I'_{21}. In the second two cases the manifolds B_2 and B_3 are four dimensional manifolds locally coordinatized by x'^1, x'^2, z' and z_1' and x''^1, x''^2, z'' and z_2'', respectively, where $\pi_2^* x'^1 = x^1$, $\pi_2^* x'^2 = x^2$, $\pi_2^* z' = z$, $\pi_2^* z_1' = z_1$ and $\pi_3^* x''^1 = x^1$, $\pi_3^* x''^2 = x^2$, $\pi_3^* z'' = z$, $\pi_3^* z_2'' = z_2$. The corresponding differential ideals I_2' and I_3' are as follows:

$$I_2' = \{dz' \wedge dx'^2 - z_1' dx'^1 \wedge dx'^2, \; dz_1' \wedge dx'^1 + \sin z' \; dx'^1 \wedge dx'^2\},$$

$$I_3' = \{dz'' \wedge dx''^1 + z_2'' \; dx''^1 \wedge dx''^2, \; dz_2'' \wedge dx''^2 - \sin z'' \; dx''^1 \wedge dx''^2\}.$$

The diffeomorphism of B_2 and B_3, $\phi : B_2 \to B_3$, by

$$x''^1 = x'^2, \; x''^2 = x'^1, \; z'' = z', \; z_2'' = z_1'$$

satisfies the condition $\phi^* I_3' = I_2'$.

To make the transition to the Wahlquist–Estabrook procedure in the case where dim $N_2 = 1$ we choose the coordinates introduced above (deleting the primes) and standard coordinates y on N_2 and y', y_a' ($a = 1, 2$) on $J^1(M,N_2)$. The coordinate presentation of the map ζ is given by

$$y' = y$$

$$y_a' = \zeta_a$$

where ζ_a denote functions on $B \times N_2$. We consider two cases:

Case 1 : Here $B = B_1 = J^1(M,N_1)$ and $I' = I_1'$.

Application of (7.5), when it is assumed that $\partial \zeta_a / \partial x^b = 0$, leads to the unique result [66]

$$\zeta_1 = 2a \sin \left(\frac{y+z}{2}\right) + z_1 \quad,$$ (7.6)

$$\zeta_2 = 2a^{-1} \sin\left(\frac{y-z}{2}\right) - z_2 .$$

The corresponding Bäcklund map determines a Bäcklund self-transformation (cf. section 6) and is the Bäcklund map of example 3.1.

Case 2 : Here $B = B_3$ and $I' = I_3'$.

Application of (7.5), when it is assumed that $\partial \zeta_a / \partial x^b = 0$, cannot lead to a Bäcklund self-transformation. It does lead to the result:

$$\zeta_1 = \tfrac{1}{2} a(1 - y^2)\sin z + a y \cos z \quad,$$ (7.7)

$$\zeta_2 = a^{-1}y - (1 + y^2) . \tfrac{1}{2} z_2 .$$

The corresponding Bäcklund map, ψ, given in example (3.2), cannot be related to a Bäcklund map ψ^ϕ, which determines a Bäcklund self-transformation, by the diffeomorphisms considered in section 6. Such diffeomorphisms can be used, however, to express the Bäcklund-related equation in a simple form. If $\phi = \mathrm{id}_M$, $\phi_1^{\;o} = \mathrm{id}_{J^o(M,N_1)}$ and $\phi_2^{\;o} : J^o(M,N_2) \to J^o(M,N_2)$ is given by $y \mapsto v = 2 \tan^{-1} y$, then the Bäcklund map ψ^ϕ is determined by the functions

$$\psi_1^\phi = a \sin (v + z) \quad ,$$

$$\psi_2^\phi = \frac{1}{a} \sin v - z_2 \quad .$$

This map determines a Bäcklund transformation between Z given by $z_{12} - \sin z = 0$ and Z' given by

$$v_{12} - (1 - (v_2/a)^2)^{\frac{1}{2}} \sin v = 0 \tag{7.8}$$

(cf. Example 3.2).   □

Example 7.2. The Korteweg-deVries equation. In the first part of this example we show how the Wahlquist-Estabrook procedure for the KdV equation [72] may be recovered from the jet bundle formulation of the Bäcklund problem. In the second part, in which we assume some familiarity with reference [72], we describe some extensions of the results of Wahlquist and Estabrook. A more complete account of some of these may be found in [67].

Since the KdV equation is of third order, one would expect to be able to solve the Bäcklund problem by a map

$$\psi: J^2(M,N_1) \times N_2 \to J^1(M,N_2) \quad .$$

For the KdV equation, which is invariant under translations of M, we make the simplifying assumption that the functions ψ_b^A in (4.6) do not depend on the coordinates x^a. It remains to determine these functions so that the integrability conditions (4.9) are equivalent to the KdV equation (4.42). A calculation now leads to the conclusion that, in the notation of examples

113

4.1 and 4.2, these functions must be of the form

$$\psi^A_1 = B^A + z_{12} \, C^A \tag{7.9}$$

$$\psi^A_2 = A^A + z_{12} \, \partial B^A/\partial z_{11} + \tfrac{1}{2}(z_{12})^2 \, \partial C^A/\partial z_{11} + z_{22} \, C^A \tag{7.10}$$

where A^A, B^A and C^A are functions of z, z_1, z_2, z_{11} and the y^A only, which satisfy the system of partial differential equations

$$z_1 \partial C^A/\partial z + z_{11} \partial C^A/\partial z_1 - \partial B^A/\partial z_2 + [B,C]^A = (12zz_1 + z_2) \partial C^A/\partial z_{11},$$

$$\partial C^A/\partial z_{11} - \partial^2 B^A/\partial z_{11} \, \partial z_2 + [\partial B/\partial z_{11}, \, C]^A = 0,$$

$$\partial^2 C^A/\partial z_{11} \, \partial z_2 + [C, \, \partial C/\partial z_{11}]^A = 0,$$

$$z_1 \, \partial^2 B^A/\partial z \, \partial z_{11} + z_{11} \, \partial^2 B^A/\partial z_1 \partial z_{11} + \partial A^A/\partial z_2 \tag{7.11}$$

$$- z_2 \, \partial C^A/\partial z - \partial B^A/\partial z_{11} + [B, \, \partial B/\partial z_{11}]^A + [C,A]^A$$

$$= (12zz_1 + z_2) \, \partial^2 B^A/\partial(z_{11})^2 \,,$$

$$z_1 \, \partial A^A/\partial z + z_{11} \, \partial A^A/\partial z_1 - z_2 \, \partial B^A/\partial z + [B,A]^A$$

$$= (12zz_1 + z_2) \, \partial A/\partial z_{11}.$$

Here $[B,C]^A$ denotes $B^B \, \partial C^A/\partial y^B - C^B \, \partial B^A/\partial y^B$, and so on.

Wahlquist and Estabrook choose the effective ideal generated by $(dz - z_1 dx^1) \wedge dx^2$, $(dz_1 - z_{11} dx^1) \wedge dx^2$, and $dz \wedge dx^1 - (dz_{11} + 12z_1 z dx^1) \wedge dx^2$. Putting this into the jet bundle context we identify it as an ideal I_2' on $J^2(M,N_1)$; its characteristic distribution is generated by $\{\dfrac{\partial}{\partial z_2} , \dfrac{\partial}{\partial z_{12}} , \dfrac{\partial}{\partial z_{22}}\}$

so that B is a 5-dimensional manifold with local coordinates x'^1, x'^2, z', z'_1, z'_{11} determined by $\pi^* x'^1 = x^1$, ..., $\pi^* z'_{11} = z_{11}$. Consequently the map ζ cannot depend on z_2, z_{12} or z_{22}, which implies that conditions (7.5) for this case may be obtained from (7.11) by imposing the further conditions

$$\partial A^A/\partial z_2 = \partial B^A/\partial z_2 = \partial B^A/\partial z_{11} = C^A = 0. \qquad (7.12)$$

If these are introduced into (7.11) then what remain are exactly Wahlquist and Estabrook's equation (31), with the transcription of notation $(z, z_1, z_{11}, A, B) \mapsto (u, z, p, -G, -F)$. They are able to integrate completely what remains, obtaining, in our notation

$$A^A = 2(z_{11} + 6z^2)X_2^{\ A} + 3(2zz_{11} + 8z^3 - z_1^{\ 2})X_3^{\ A}$$

$$- 8X_4^{\ A} - 8zX_5^{\ A} - 4z^2X_6^{\ A} - 4z_1X_7^{\ A} , \qquad (7.13)$$

$$B^A = - 2X_1^{\ A} - 2zX_2^{\ A} - 3z^2X_3^{\ A} ,$$

where the $X_\alpha^{\ A}$ ($\alpha = 1, \ldots, 7$) are functions $X_\alpha^{\ A}(y)$ on N_2, and the corresponding vector fields $X_\alpha = X_\alpha^{\ A}\partial/\partial y^A$ must satisfy the conditions

$$[X_1, X_2] = - X_7, \quad [X_1, X_7] = X_5, \quad [X_2, X_7] = X_6 , \qquad (7.14)$$

$$[X_1, X_3] = [X_2, X_3] = [X_1, X_4] = [X_2, X_6] = 0 ,$$

$$[X_1, X_5] + [X_2, X_4] = 0, \quad [X_3, X_4] + [X_1, X_6] + X_7 = 0. \qquad (7.15)$$

Taking (7.13) into account and comparing (7.9) with (5.31),

one concludes that the factorization required for the existence
of a structure group is here satisfied automatically.

Notice that this calculation leaves unspecified dim N_2,
which is the range of the index A. Estabrook and Wahlquist leave
open the question of structure of the collection of vector fields
X_α, remarking that "this structure comes close to defining a Lie
algebra" and that "the process is apparently open-ended; the
whole process does not appear to close itself off into a unique
Lie algebra with any finite number of generators..." (that is,
with a finite basis). They proceed to "force closure by
arbitrarily imposing linear dependence among the generators".

From our point of view, the situation is the following: A
check of the calculations in [72] shows that conditions (7.14)
and (7.15) have different status; equations (7.14) are definitions of X_5, X_6 and X_7 in terms of X_1 and X_2, while (7.15) are
relations between commutators and iterated commutators of X_1, X_2,
X_3 and X_4. Thus one is led to consider the Lie algebra L generated by X_1, X_2, X_3 and X_4 subject to the relations (7.15). This
algebra may be shown to be infinite-dimensional [67]. The finite-
dimensional Lie algebras which appear in solutions of the
Bäcklund problem are homomorphic images of L . When Wahlquist and
Estabrook "force closure", they in effect choose a homomorphism
by defining its kernel. In [72], they set

$$X_9 = \lambda(X_7 - X_8)$$

where

$$X_9 := [X_1, X_5], \qquad X_8 := [X_4, X_3],$$

and λ is an arbitrary parameter. With the help of the Jacobi identities and (7.15) they can now calculate the whole multiplication table for the Lie algebra $\hat{L}$ isomorphic to L/K, where K is the ideal in L generated by $X_9 - \lambda(X_7 - X_8)$.

It is useful to decompose $\hat{L}$ into the sum of its radical and a semi-simple sub-algebra. It happens that in this case the sum is direct. In working out the decomposition we exploited the theorem that the radical of a Lie algebra is the orthogonal complement (by the Killing form) of the first derived ideal ([47] p.73). We find that

$$\hat{L} = SL(2,\mathbb{R}) \oplus H \tag{7.17}$$

where $\oplus$ denotes the direct sum of Lie algebras and H is a five-dimensional nilpotent Lie algebra with multiplication table, in a basis $H_1, \ldots, H_5$, given by

$$[H_1, H_2] = [H_3, H_4] = H_5 \tag{7.18}$$

and other brackets all zero.

A change of basis which exhibits the decomposition explicitly is given by

$$A_1 = -X_5, \quad A_2 = X_5 + \lambda X_6, \quad A_3 = X_7 - X_8, \quad H_1 = X_3,$$

$$H_2 = X_4 - \lambda X_5 - \lambda^2 X_6, \quad H_3 = X_1 + X_5 + \lambda X_2, \quad (7.19)$$

$$H_4 = X_2 - X_6, \quad H_5 = -X_8.$$

In this basis A_1, A_2 and A_3 generate $SL(2,\mathbb{R})$, while the H's are as described above, and all A's commute with all H's.

We remark without giving details that a different choice of kernel yields the 11-dimensional Lie algebra $SL(2,\mathbb{R}) \oplus SL(2,\mathbb{R}) \oplus H$, and we conjecture that a $(3n + 5)$-dimensional Lie algebra $\oplus_n SL(2,\mathbb{R}) \oplus H$ can be achieved in a similar way.

The parameter λ may be seen to arise from the Galilean transformation of example 6.3: this transformation and its prolongation to B induce an automorphism ϕ_G of L [67] given by

$$\phi_G : X_1 \mapsto X_1 + tX_2 + \frac{3}{2} t^2 X_3$$

$$X_2 \mapsto X_2 + 3tX_3$$

$$X_3 \mapsto X_3 \quad (7.20)$$

$$X_4 \mapsto X_4 + 3tX_1 + \frac{3}{2} t^2 X_2 + \frac{3}{2} t^3 X_3 + tX_5 + \frac{1}{2} t^2 X_6 .$$

The algebra L/K is naturally isomorphic to $\phi_G(L)/\phi_G(K)$ by $X + K \mapsto \phi_G X + \phi_G(K)$ and it is easily verified from (7.20) that $\phi_G(K)$ is the ideal in L generated by $X_9 - (\lambda - 2t)(X_7 - X_8)$. Thus the Galilean transformatión gives rise to a 1-parameter

family of 8-dimensional algebras - all isomorphic to
$SL(2,\mathbb{R}) \oplus H$.

The proof that L is infinite dimensional, given in [67],
exploits the automorphism ϕ_G, together with another automorphism
arising from the scale-invariance of the KdV equation. With the
help of these automorphisms it is shown also that the 5-dimensional
algebra H is closely related to the classical symmetries and
conservation laws of the KdV equation.

In their work Estabrook and Wahlquist choose a non-linear
eight-dimensional representation of $\hat{L}$ by vector fields, and
proceed to recover various results concerning the KdV equations,
including the equations of the inverse scattering problem, which
were identified by Crampin and two of us [18] as equations of
parallel transport by an $SL(2,\mathbb{R})$-connection. In the present
version the connection enters as soon as one constructs a map
$f : M \rightarrow N_1$, as explained earlier; our previous version employed
the machinery of the principal bundle, which from the point of
view of this paper is unnecessary baggage - a principal bundle
may always be associated to a fibre bundle with structural
group [23], but this does not seem particularly advantageous to
the formalism developed above. □

We conclude this section by explaining in the jet bundle
context a method developed by Wahlquist and Estabrook [72] for
construction of Bäcklund automorphisms from Bäcklund maps which
are not automorphisms. This method depends on a further

generalization of the notion of prolongation introduced in section 2 and extended in section 6.

First we introduce the fibred product: as usual, M, N_1 and N_2 are manifolds, and

$$J^\ell(M,N_1) \times_M J^k(M,N_2) = \{(\xi_1,\xi_2) \in J^\ell(M,N_1) \times J^k(M,N_2) \mid \alpha(\xi_1) = \alpha(\xi_2)\}.$$

In particular

$$J^\ell(M,N_1) \times_M J^\ell(M,N_2) \cong J^\ell(M,N_1 \times N_2)$$

is an isomorphism [31].

For simplicity, we consider only diffeomorphisms which leave M and N_1 pointwise fixed. Thus we suppose given (or to be sought for) a diffeomorphism

$$\tilde{\chi}^o : J^o(M,N_1 \times N_2) \to J^o(M, N_1 \times N_2) ,$$

compatible with the identity map on $J^o(M,N_1)$. If x^a, z^μ, y^A are standard local coordinates on the domain of $\tilde{\chi}^o$, x'^a, z'^μ, y'^A on its codomain, then $\tilde{\chi}^o$ has the coordinate presentation

$$x'^a = x^a, \quad z'^\mu = z^\mu, \quad y'^A = \zeta^A(x^a, z^\mu, y^B) .$$

Let p_2 denote the natural projection

$$J^o(M,N_1 \times N_2) \cong J^o(M,N_1) \times_M J^o(M,N_2) \to J^o(M,N_2). \quad \text{Then}$$

$$\chi^o := p_2 \circ \tilde{\chi}^o : J^o(M,N_1 \times N_2) \to J^o(M,N_2)$$

also leaves M pointwise fixed.

Now $\tilde{\chi}^o$ is compatible with id_M (see section 6), hence may be

120

prolonged, according to (6.3), to a unique map

$J^{\ell}(M,N_1 \times N_2) \to J^{\ell}(M, N_1 \times N_2)$, which leaves $J^{\ell}(M,N_1)$ pointwise

fixed, since $\widetilde{\chi}^0$ leaves $J^0(M,N_1)$ pointwise fixed. Therefore this

map passes to the quotient and projects to define a unique map

$$\widetilde{\chi}^{\ell} : J^{\ell}(M,N_1) \times N_2 \to J^{\ell}(M,N_1) \times N_2$$

which leaves $J^{\ell}(M,N_1)$ pointwise fixed. The local coordinate

presentation of $\widetilde{\chi}^{\ell}$ is

$$x'^a = x^a, \quad z'^{\mu} = z^{\mu}, \quad z'^{\mu}_a = z^{\mu}_a, \quad \ldots, \quad z'^{\mu}_{a_1 \ldots a_{\ell}} = z^{\mu}_{a_1 \ldots a_{\ell}},$$

$$y'^A = \zeta^A(x^a, z^{\mu}, y^B)$$

where unprimed and primed coordinates are standard local co-

ordinates on the domain and codomain of $\widetilde{\chi}^{\ell}$ respectively. More-

over $\chi^{0^{\cdot}}$ may be prolonged to a unique map

$$\chi^{\ell} : J^{\ell}(M, N_1 \times N_2) \to J^{\ell}(M, N_2)$$

by (6.3).

Now let $\psi : J^h(M,N_1) \times N_2 \to J^1(M,N_2)$ be a given ordinary

Bäcklund map with integrability conditions Z on $J^{h+1}(M,N_1) \times N_2$.

Then the map

$$((\pi^h_1 \circ pr_1) \times \psi) \circ \Delta(J^h(M,N_1) \times N_2) :$$

$$J^h(M,N_1) \times N_2 \to J^1(M,N_1) \times J^1(M,N_2)$$

passes to the quotient to define a map

$$\tilde{\psi} : J^h(M,N_1) \times N_2 \rightarrow J^1(M,N_1) \times_M J^1(M,N_2) \cong J^1(M,N_1 \times N_2).$$

The coordinate presentation of $\tilde{\psi}$ is

$$x'^a = x^a, \quad z'^\mu = z^\mu, \quad z'^\mu_a = z^\mu_a, \quad y'^A = y^A, \quad y'^A_a = \psi^A_a,$$

where ψ^A_a are the functions on $J^h(M,N_1) \times N_2$ determining the Bäcklund map ψ. Finally, let

$$\psi^X := \chi^1 \circ \tilde{\psi} \circ (\tilde{\chi}^h)^{-1} : J^h(M,N_1) \times N_2 \rightarrow J^1(M,N_2).$$

With singly-primed coordinates on its domain, doubly-primed coordinates on its codomain, this map has the coordinate presentation

$$x''^a = x'^a, \quad y''^A = y'^A, \quad y''^A_a = \psi^{X^A}_a,$$

where $\psi^{X^A}_a = f^A_a \circ (\tilde{\chi}^h)^{-1}$ and f^A_a are functions on the codomain of $(\tilde{\chi}^h)^{-1}$ which in unprimed standard coordinates have the presentation

$$f^A_a = (\partial\zeta^A/\partial y^B)\psi^B_a + (\partial\zeta^A/\partial z^\mu)z^\mu_a + \partial\zeta^A/\partial x^a .$$

Moreover, a straightforward calculation shows that

$$\psi^{X^*} \Omega^1(M,N_2) \subset \tilde{\Omega}^{\psi,h}$$

(compare section 4) whence it follows immediately that ψ^X, like ψ, solves the Bäcklund problem for the system Z.

In general, ψ^X will be distinct from ψ. It is this that makes it possible to construct Bäcklund automorphisms by the

122

above procedure.

Example 7.3: The sine-Gordon equation. Here we exhibit a diffeomorphism which relates a Bäcklund map ψ^ϕ, which does not determine a Bäcklund self-transformation of the sine-Gordon equation, to the Bäcklund automorphism, ψ^X, for this equation (cf. examples 3.1, 3.2 and 7.1).

Using primed coordinates we write the functions ψ_1, ψ_2, determining the Bäcklund map $\psi : J^1(M,N_1) \times N_2 \rightarrow J^1(M,N_2)$ of example 3.2 (the map ψ^ϕ of example 7.1) in the form

$$\psi_1 = a \sin(y' + z')$$

$$\psi_2 = a^{-1} \sin y' - z'_2 .$$

Let $\tilde{\chi}^o : J^0(M,N_1 \times N_2) \rightarrow J^0(M, N_1 \times N_2)$ be given by

$$x'^a = x^a$$

$$z' = z$$

$$y' = z + 2y .$$

The map $\psi^X : J^1(M,N_1) \times N_2 \rightarrow J^1(M,N_2)$, constructed as above, is determined by the functions

$$\psi_1{}^X = 2a \sin\tfrac{1}{2}(y + z) + z_1 ,$$

$$\psi_2{}^X = 2a^{-1} \sin\tfrac{1}{2}(y - z) - z_2 ,$$

and is the Bäcklund automorphism for the sine-Gordon equation [10].

□

REFERENCES

[1]   Ablowitz, M.J., Kaup, D.W., Newell, A.C. and Segur, H. :
      1973, Phys. Rev. Lett. 31, pp. 125-127.

[2]   Ablowitz, M.J., Kaup, D.W., Newell, A.C. and Segur, H. :
      1974, Stud. Appl. Math. 53, pp. 249-315.

[3]   Abraham, R. : 1963, *Lectures of Smale on differential
      topology*.

[4]   Backlund, A.V. : 1875, Lund Univ. Arsskrift, 10.

[5]   Backlund, A.V. : 1883. Lund Univ. Arsskrift, 19.

[6]   Brickell, F. and Clark, R.S. : 1970, *Differentiable
      manifolds*, Van Nostrand Reinhold, London.

[7]   Calogero, F. and Degasperis, A. : 1976, Nuovo Cimento 32B,
      pp. 201-242.

[8]   Cartan, É. : 1946, *Les systèmes différentielles
      extérieures et leurs applications géométriques*, Hermann,
      Paris.

[9]   Chaichian, M. and Kulish, P.P. : 1978, Phys. Lett. 78B,
      pp. 413-416.

[10] Chen, H. : 1974, Phys. Rev. Lett. 33, pp. 925-928.

[11] Clairin, J. : 1902, Ann. Scient. Éc. Norm. Sup., 3e sér.
      Supple. 19, pp. S1-S63.

[12] Corones, J.P. : 1977, J. Math. Phys. 18, pp. 163-164.

[13] Corones, J.P., Markovski, B.L., and Rizov, V.A. : 1977,
      J. Math. Phys. 18, pp. 2207-2213.

[14] Corrigan, E.F., Fairlie, D.B., Yates, R.G. and Goddard, P. :

1978, Commun. Math. Phys. 58, pp. 223-240.

[15] Crampin, M. : Phys. Lett. 66A, pp. 170-172.

[16] Crampin, M., Hodgkin, L. McCarthy, P. and Robinson, D.C. : 1979, Rep. Mathematical Phys., to appear.

[17] Crampin, M. and McCarthy, P. : 1978, Lett. Math. Phys. 2, pp. 303-312.

[18] Crampin, M., Pirani, F.A.E. and Robinson, D.C. : 1977, Lett. Math. Phys. 2, pp. 15-19.

[19] Dhooghe, P.F.J. : 1978, Jet bundles and Bäcklund transformations. In press.

[20] Dhooghe, P.F.J. : 1978, Le problème de Bäcklund, Preprint.

[21] Dodd, R.K. and Gibbon, J.D. : 1978, Proc. R. Soc. Lond. A 359, pp. 411-433.

[22] Ehresmann, C. : 1951, In *Colloque de topologie (espaces fibrés)*, Masson, Paris.

[23] Ehresmann, C. : 1953, Introduction a la théorie des structures infinitésimales et des pseudo-groupes de Lie. Colloque Internationale de C.N.R.S., Paris.

[24] Eisenhart, L.P. : 1960, *A treatise on the differential geometry of curves and surfaces*, Dover, New York.

[25] Estabrook, F.B. and Wahlquist, H.D. : 1976, J. Math. Phys. 17, pp. 1293-1297.

[26] Forsyth, A.R. : 1906, *Theory of differential equations*, vol. 6, Cambridge University Press.

[27] Forsyth, A.R. : 1921, *A treatise on differential equations*, MacMillan, London.

[28] Gardner, C.S., Greene, J.M., Kruskal, M.D. and Miura, R.M. : 1967, Phys. Rev. Lett. 19, pp. 1095-1097.

[29] Gardner, C.S., Greene, J.M., Kruskal, M.D. and Miura, R.M. : 1974, Comm. Pure App. Math. 27, pp. 97-133.

[30] Gardner, R.B. : 1977, Lecture at the Berlin conference on Differential Geometry and Global Analysis.

[31] Goldschmidt, H. : 1967, Ann. of Math. 82, pp. 246-270.

[32] Goldschmidt, H. : 1967, J. Differential Geometry 1, pp. 269-307.

[33] Golubitsky, M. and Guillemin, V. : 1973, *Stable mappings and their singularities*, Chapter II, §2, Springer-Verlag, New York, Heidelberg, Berlin.

[34] Goursat, E. : 1925, *Le problème de Bäcklund.* Mémor. Sci. Math. Fasc. 6, Gauthier-Villars, Paris.

[35] Guillemin, V. and Sternberg, S. : 1966, Mem. Amer. Math. Soc. 64.

[36] Harrison, B.K. 1978 : Phys. Rev. Lett. 41, pp. 1197-1200.

[37] Harrison, B.K. and Estabrook, F.B. : 1971, J. Math. Phys. 12, pp. 653-666.

[38] Hermann, R. : 1965, Advances in Math. 1, pp. 265-317.

[39] Hermann, R. : 1968, *Differential geometry and the calculus of variations.* Academic Press, New York, London.

[40] Hermann, R. : 1970, *Vector bundles in mathematical physics,* Vol. 1, Benjamin, New York.

[41] Hermann, R. : 1973, *Geometry, physics and systems,* Marcel Dekker Inc., New York.

[42] Hermann, R. : 1975, *Gauge field and Cartan-Ehresmann connections,* Part A, Interdisciplinary Mathematics, Vol. X, Math. Sci. Press, Brookline Ma.

[43] Hermann, R. : 1976, Phys. Rev. Lett. 36, pp. 835-836.

[44] Hermann, R. : 1976, *The geometry of non-linear differential equations, Bäcklund transformations and solitons,* Part A, Interdisciplinary Mathematics, Vol. XII, Math. Sci. Press, Brookline Ma.

[45] Hirota, R. : 1971, Phys. Rev. Lett. 27, pp. 1192-1194.

[46] Ibragimov, N.H. and Anderson, R.L. : 1977, J. Math. Anal. Appl. 59, pp. 145-162.

[47] Jacobson, N. : 1962, *Lie algebras,* Interscience, New York, London.

[48] Johnson, H.H. : 1962, Math. Annalen 148, pp. 308-329.

[49] Korteweg, D.J., and deVries, G. : 1895, Phil. Mag. 39, pp. 422-443.

[50] Krupka, D.J. : 1975, J. Math. Anal. App. 49, pp. 180-206 and 469-476.

[51] Krupka, D.J., and Trautman, A. : 1974, Bull. Acad. Pol. Sci., Sér. sci. math., astro. phys., pp. 207-211.

[52] Lamb, G.L. : 1976, In *Bäcklund transformations* (ed. R.M. Miura) pp. 69-79, Lecture Notes in Mathematics, Vol. 515, Springer-Verlag, Berlin, Heidelberg, New York.

[53] Lang, S. : 1965, *Algebra*, Addison-Wesley, Reading, Ma.

[54] Lax, P.D. : 1968, Comm. Pure. App. Math. XXI, pp. 467-490.

[55] Lie, S. : 1879, Ark. for Math. og Nat. 4, p.150.

[56] Loewner, C. : 1950, NACA Tech. Note 2065.

[57] Lychagin, V.V. : 1975, Russian Math. Surveys 30, pp. 105-175.

[58] McCarthy, P. : 1978, Lett. Math. Phys. 2, pp. 167-170.

[59] Morris, H.C. : 1977, Int. J. Theo. Phys. 16, pp. 227-231.

[60] Morris, H.C. : 1977, J. Math. Phys. 18, pp. 530-532.

[61] Morris, H.C. : 1977, J. Math. Phys. 18, pp. 533-536.

[62] Nelson, E. : 1967, *Tensor analysis*, Princeton University Press.

[63] Pirani, F.A.E. and Robinson, D.C. : 1977, C.R. Acad. Sci. Paris 285, pp. 581-583.

[64] Rogers, C. : 1976, In *Bäcklund transformations* (ed. R.M. Miura) pp. 106-135, Lecture Notes in Mathematics, Vol. 515, Springer-Verlag, Berlin, Heidelberg, New York.

[65] Scott, A.C. Chu, F.Y.F. and McLaughlin, D.W. : 1973, Proc. I.E.E.E. 61, pp. 1443-1483.

[66] Shadwick, W.F. : 1978, J. Math. Phys. 19, pp. 2312-2317.

[67] Shadwick, W.F. : 1978, The KdV prolongation algebra. Preprint.

[68] Steudel, H. : 1975, Ann. Physik (7) 32, pp. 445-455.

[69] Su, C.H. and Gardner, C.S. : 1969, J. Math. Phys. 10, pp. 536-539.

[70] Trautman, A. : 1972, In *General relativity* (ed. L. O'Raifeartaigh) pp. 85-99, Clarendon Press, Oxford.

[71] Wahlquist, H.D. and Estabrook, F.B. : 1973, Phys. Rev. Lett. 31, pp. 1386-1390.

[72] Wahlquist, H.D. and Estabrook, F.B. : 1975, J. Math. Phys. 16, pp. 1-7.

[73] Zakharov, V.E. and Shabat, A.B. : 1972, Sov. Phys. JETP 34, pp. 62-69.

INDEX